# FUNDAMENTALS OF TROPICAL TURF MANAGEMENT

# Dedication

I dedicate this book to the most beloved people in my life – my wife Alicja and our son Filip

# Fundamentals of Tropical Turf Management

**G. Wiecko**
*University of Guam CNAS*
*Mangilao*
*USA*

www.cabi.org

**CABI is a trading name of CAB International**

CABI Head Office
Nosworthy Way
Wallingford
Oxfordshire OX10 8DE
UK

Tel: +44 (0)1491 832111
Fax: +44 (0)1491 833508
E-mail: cabi@cabi.org
Website: www.cabi.org

CABI North American Office
875 Massachusetts Avenue
7th Floor
Cambridge, MA 02139
USA

Tel: +1 617 395 4056
Fax: +1 617 354 6875
E-mail: cabi-nao@cabi.org



A catalogue record for this book is available from the British Library, London, UK.

**Library of Congress Cataloging-in-Publication Data**

Wiecko, G. (Greg)
Fundamentals of tropical turf management / by G. Wiecko.
p.cm.
Includes bibliographical references (p. ) and index.
ISBN-13: 978-1-84593-030-1 (alk. paper)
ISBN-10: 1-84593-030-4(alk. paper)
1. Turf management. 2. Turfgrasses. I. Title.

SB433.W48 2006
635.9′642--dc22

2005014356

ISBN-10: 1-84593-030-4
ISBN-13: 978-1-84593-030-1

Typeset by SPi, Pondicherry, India
Printed and bound in the UK by Biddles Ltd, King's Lynn.

# Contents

# Preface

This book is designed to serve as a basic text for beginning students in the fields of turf-grass management, horticulture, landscape architecture and agronomy. It is written for students around the world but should be especially helpful to those in tropical regions of south-east Asia, Central America and the coastal regions of eastern and western Africa. The book is intended to present basic turf-grass management topics in a simple and direct way that can be easily understood by readers from diverse cultures. The large number of illustrations presented throughout the text is intended to supplement reading as well as to help explain key concepts and relationships between turf-grass plants and their environment. Most of the information presented in the text was drawn from published texts, from my experience and from discussions with experienced practitioners. I am very grateful to Ric Castro, who produced most of the illustrations in the volume, and to Anne B. Thistle for her valuable editorial help. My most sincere appreciation is due to the University of Guam Endowment Foundation for helping with expenses associated with preparation of this book as well as to everyone who contributed to successful completion of the work.

Greg Wiecko
*University of Guam*

# 1 Introduction to Turf-grasses

Turf-grasses are plants of the family *Poaceae* that form a more or less continuous ground cover that persists under regular mowing and traffic. Turf, on the other hand, is the community of turf-grass plants and the soil attached to their roots. Turf-grass therefore refers only to the plant community, whereas turf represents a higher level of ecological organization by including the soil (or any other medium in which the turf-grasses are grown) and its resident organisms. For example, turf around a tropical resort consists of many turf-grasses selected for recreational, athletic and ornamental purposes.

Most of the turf-grasses we currently use were developed under the selection pressure of grazing animals. Before mechanical mowers were readily available, sheep grazing was used to maintain golf courses and other recreational sites. When grasses were subjected to grazing, only those plants with their growing points near the soil surface could survive (Fig. 1.1). These were the first grasses we could call turf-grasses. They formed a continuous cover and could persist under regular mowing or grazing. Advances in the development of mechanical mowers allowed turf-grasses to be cut shorter and resulted in further selection for low growing points. Now, some turf-grasses can be mowed as close as 3–4 mm above the ground, forming turf areas as thick, firm and smooth as carpets.

Turf plays many roles in our lives. Its dense growth and extensive root system prevent soil erosion by wind and water, it removes smoke and dust from the atmosphere, and it releases oxygen into the air. Its uniform green colour reduces visual pollution, especially along

**Fig. 1.1.** In the past, grazing sheep and goats were used to maintain grassy areas.

roadsides, and it reduces glare, muffles sound (particularly important in noisy cities) and provides a cooling effect. Turf also provides a soft cushion on playgrounds and athletic fields. Turf can be grown in almost any location. Its growth is perennial, and it provides a green cover that does not need to be replaced if properly managed. Turf is relatively easy to grow and quite forgiving of occasional mismanagement.

The most basic function of any turf is soil stabilization (Fig. 1.2). Soil covered by turf does not slide, does not flow with water, and does not cling to feet or shoes. Turf also prevents or reduces the formation of water puddles.

In residential and commercial lawns, turf's function is primarily aesthetic. Humans have a natural desire to surround themselves with beauty. Plants, including turf-grasses, fulfil a large part of this desire (Fig. 1.3).

Finally, the turf used in stadiums and on other athletic fields must possess different qualities. It must withstand much more extreme physical stresses than turfs not associated with sports and recreational activities (Fig. 1.4).

Turfs used for stabilization, ornamental and recreational purposes require different intensity of maintenance. For example, turfs used exclusively for soil stabilization, called utility turfs, require only sporadic mowing, little or no fertilization or irrigation, and no additional cultural

**Fig. 1.2.** Established turf on the steep slope effectively prevents soil erosion.

**Fig. 1.3.** A home lawn in Asia.

**Fig. 1.4.** Sport turfs must withstand extreme physical stresses.

practices (activities conducted by the turf manager to promote turf health). On the other hand, residential turfs, turfs in parks or those around businesses require more maintenance. Sports turfs usually require the highest level of care.

Some turf areas are quite uniform and others more complex. For example, athletic fields are rather uniform and often consist of a single species of turf-grass under a specific level of care. Turf on these areas, besides its pleasing appearance, must satisfy requirements associated with a particular sport, be able to withstand the physical stresses it imposes and have the capacity to recover from injury between uses. On the other hand, golf course turf is a mixture of different grasses. It may include different species or varieties; they may have different appearances, serve different functions and need to be maintained at different levels of intensity. Greens and tees are always the most intensively maintained, followed by fairways, and then roughs, which are often maintained as utility turfs.

Regardless of the particular use to which turfs are subjected and the intensity of culture applied, they are incredibly complex systems. In these ecosystems, turf-grass communities exist in close association with climate, soil and biotic environments. The quality and persistence of turfs reflect how well all of the components interact to form a sustainable and efficient association.

## Climate Adaptation

Grasses are divided into two major groups of species: (i) those suitable for warm; and (ii) those suitable for cool climates. The first, called warm-season grasses, are best adapted to temperatures around 30°C and differ from the second, called cool-season grasses, not only in climatic requirements but also in the way they carry out photosynthesis. Most of the warm-season species belong to a classification called C4 plants, in which photosynthesis is much more efficient than that in C3, or temperate-climate, plants. In addition, they use only one-third to one-half as much water to produce the same amount of dry matter, and overall they thrive in hot weather. Another unique characteristic of warm-season grasses is that they undergo dormancy when soil temperature drops below 10°C. They lose their chlorophyll and remain yellow or brown until the temperature rises above 10°C again. Tropical grasses are essentially warm-season grasses that never become dormant and often manifest little tolerance for cold stress. This volume will discuss only warm-season species grown in the climates where temperature is relatively uniform and warm year-round.

Turf-grasses, like all other plants, undergo physiological processes that govern their growth and development. Turf physiology is very complex and includes thousands of such processes. The most essential are photosynthesis, respiration, growth and storage.

## Physiology

### Photosynthesis

Plants have the unique ability to change carbon dioxide from the air and water from the soil into carbohydrates, such as glucose and oxygen. This conversion process, called photosynthesis, takes place in the leaves and green stems and requires energy supplied by the sun. Carbohydrates are used by the plant to make energy needed for its growth and development, and oxygen is released into the air as a by-product. In general, ensuring maximum photosynthesis requires maintaining the largest possible leaf area; in other words, it is physiologically beneficial to maintain turf at the highest possible mowing height typical for the given species.

Photosynthesis is greatest in bright sunlight. Tropical, warm-season grasses, unlike some cool-season grasses, have almost unlimited photosynthetic potential. High temperature is seldom a limiting factor either, as long as the plant has adequate moisture. On the other hand, warm-season grasses are less efficient at photosynthesis under low-light conditions or at cool temperatures than are grasses from cooler climates. In many cases, shade results in poor growth, and when the temperature drops to around 10°C, photosynthesis often comes to a complete stop, and the grass undergoes dormancy. In the tropics these conditions may be encountered at higher elevations (Fig. 1.5).

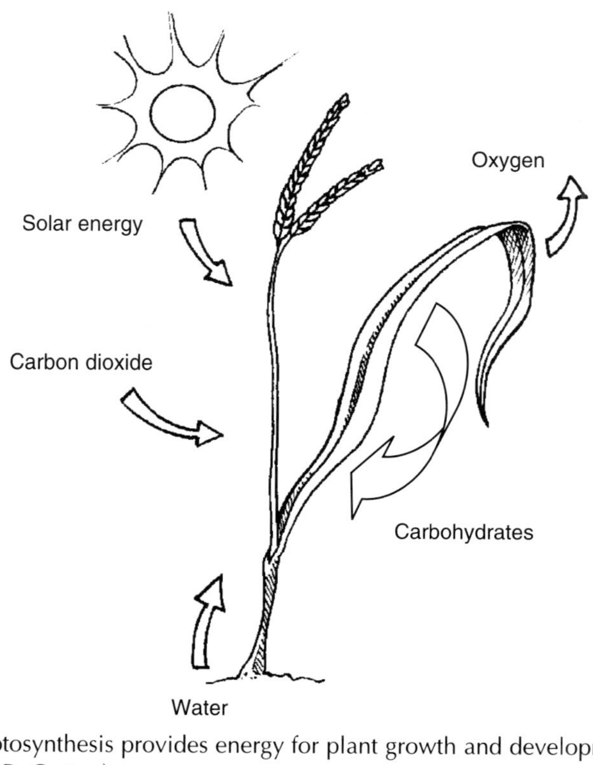


**Fig. 1.5.** Photosynthesis provides energy for plant growth and development. (Drawing by R. Castro.)

## Respiration

During darkness, turf plants do not produce carbohydrates but still need some energy to maintain essential functions. Although photosynthesis stops at night, respiration continues (Fig. 1.6). Carbohydrates stored during the day are broken down during the night by respiration, which releases energy, water and carbon dioxide. High temperatures increase respiration, especially at night. Warm-season grasses are adapted to these conditions and can tolerate hot days as well as nights without negative effects. Under the same conditions, cool-season grasses may suffer severe harm. Respiration also increases when photosynthesis is reduced, e.g. during periods of drought or after mowing.

## Growth

Growth can be described as irreversible increase in size. Grasses grow by cell division (the process in which individual cells divide in two, producing more cells) and by cell elongation (in which individual cells stretch or lengthen). Growing cells synthesize carbohydrates, proteins, fats, DNA and other nucleic acids, and many other compounds, a process that requires a sizeable amount of energy derived directly from photosynthesis. Again, efficient synthesis of these compounds by warm-season grasses is almost unlimited at high temperatures. As long as moisture and light are unrestricted, growth is unrestricted as well.

## Storage

Carbohydrates and sometimes proteins are stored for use later when needed. For example, storage products are used whenever the plant is

Photosynthesis

Carbon dioxide + Water + Solar energy → Glucose + Water + Oxygen

$$6CO_2 + 12H_2O + \text{Light} \longrightarrow C_6H_{12}O_6 + 6H_2O + 6O_2$$

Respiration

Glucose + Water + Oxygen → Carbon dioxide + Water + Chemical energy

$$C_6H_{12}O_6 + 6H_2O + 6O_2 \longrightarrow 6CO_2 + 12H_2O + \text{Chemical energy}$$

**Fig. 1.6.** Respiration is the reverse of photosynthesis. In order for the plant to grow, production of carbohydrates must exceed their breakdown.

forced to grow beyond its normal rate, as it does after application of fertilizer or after the loss of leaves to mowing. Plants must maintain reserves for the future. The level of storage reserves is a good indicator of how well a plant will survive stressful conditions, such as drought, disease infection, insect damage or excessive traffic.

The four principal physiological processes are therefore interrelated. Under high photosynthesis enough energy is captured and enough carbohydrates produced to support growth, storage and respiration. If photosynthesis decreases, stored products must be used for respiration and growth. For example, mowing turf too low (scalping) removes too much leaf area and disturbs photosynthesis, causing the turf plant to 'borrow' energy from storage to maintain respiration and to be able to regrow. Repeated scalping may deplete storage and may lead to starvation and possibly death of the plant. Any cultural practice that increases growth, such as fertilization, uses up carbohydrates that would otherwise be sent to storage. Whenever a plant is fertilized, growth is stimulated and storage is reduced. Frequent fertilization may therefore harm plants with low storage reserves, and when those plants undergo stress, they may suffer great damage. Understanding how plants grow and respond to the environment is essential to maintenance of high-quality turf. Good turf management requires knowledge of basic physiological processes that take place in plants and use of that knowledge to best advantage.

## Morphology

### Shoots

As indicated in Fig. 1.7, the above-ground part of the plant is known as the *shoot*. A grass shoot consists of *leaves, stem* and *inflorescence*. Each leaf is composed of two defined parts: (i) the upper, mostly flat part, called the *blade*; and (ii) the lower, often tubular part, called the *sheath*. The swollen areas along the stem where leaves are attached are called *nodes*. The node is the point at which the vascular system of the leaf is attached to that of the stem. At each node, in the angle between the leaf and the stem, is a *bud*, from which new stem growth can originate. An abundance of buds assures continuous formation of new stems and uninterrupted growth of the grass plant. The portions of the stem between nodes are called *internodes*. Internode length in most turf-grasses ranges from a fraction of a millimetre to several centimetres. Turf density depends highly on internode length. Shorter internodes result in higher density of a turf's canopy.

Very close to ground level, numerous nodes with very short internodes are stacked on top of each other, forming the *crown* (Fig. 1.8). The

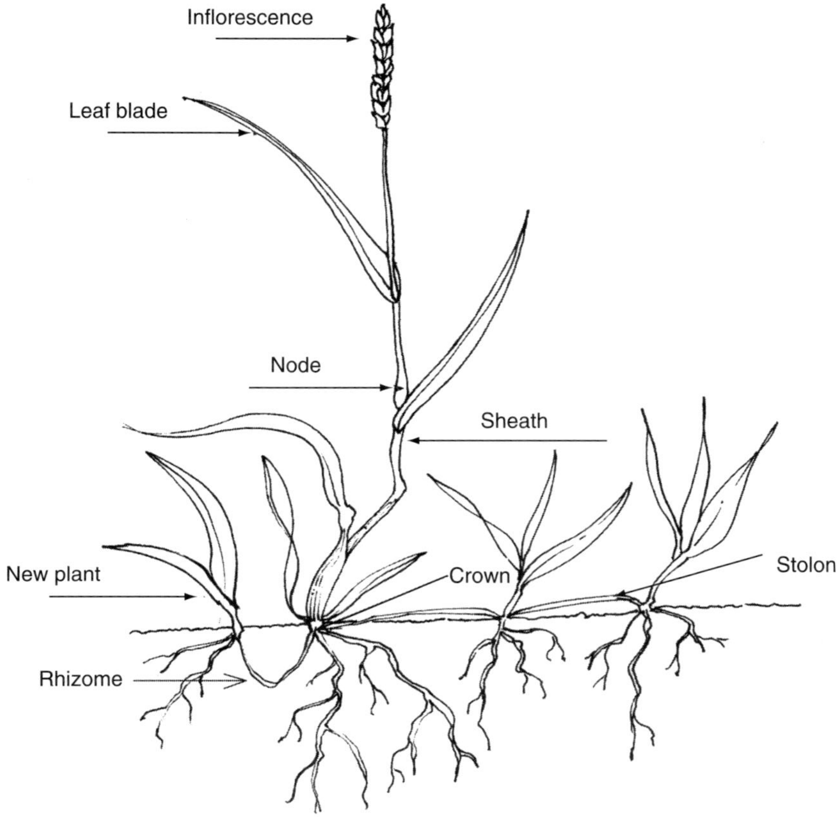


**Fig. 1.7.** Essential parts of the grass plant. (Drawing by R. Castro.)

crown is the most vital part of the grass plant. As long as the crown is alive, even profoundly injured plants may survive. Serious injury to the crown usually results in death of the plant. For example, excessively low mowing height may result in massive destruction of crowns and, as a consequence, major losses of turf. Some diseases and insects attack the crown, also causing severe damage to turf.

When new stems, usually growing from buds in the crown area, grow laterally and extend roots of their own into the soil, they can develop into new plants and survive independently if severed from the parent plant. These primary lateral shoots are called *tillers*. Tillers are typical of all turf-grass species. In addition to tillers, some species develop underground lateral stems called *rhizomes* (Fig. 1.9), and others produce above-ground lateral stems called *stolons* (Fig. 1.10). Some species develop both rhizomes and stolons. The majority of turf-grasses grown in warm climates are rhizomatous (rhizome-bearing) or stoloniferous

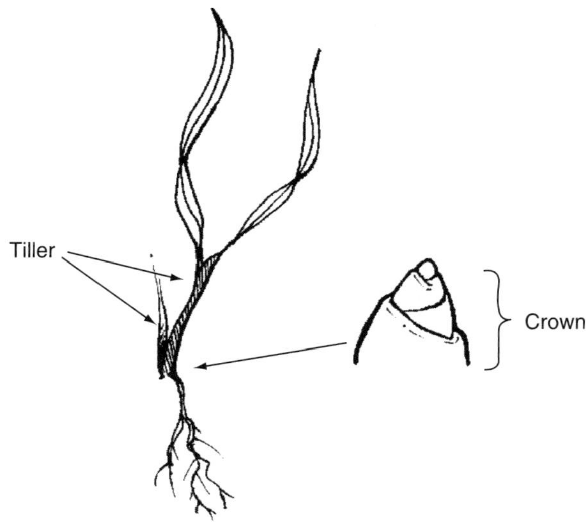


**Fig. 1.8.** Tillers emerge from the crown and, after developing roots, may become individual plants. (Drawing by R. Castro.)

(stolon-bearing) or both and can 'creep', that is, grow laterally and gradually fill up the bare soil between plants. This feature is especially important in species and/or varieties that do not produce viable seeds and must be propagated vegetatively.

Because rhizomes grow below ground, in the dark, they do not carry out photosynthesis as stolons do. Their lack of chlorophyll makes them look like large white roots. They are not roots, however, because they develop nodes with buds (which roots cannot) and, if separated from the

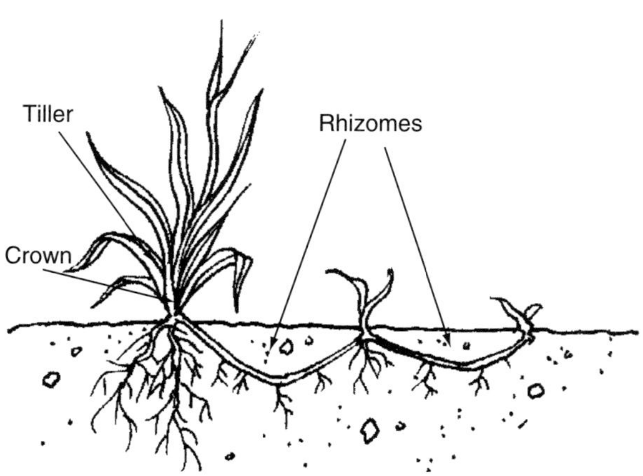


**Fig. 1.9.** Rhizomes are underground stems from which new plants can emerge. (Drawing by R. Castro.)

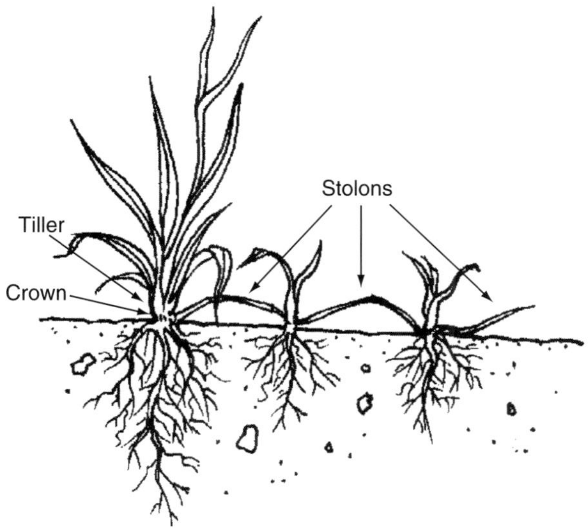


**Fig. 1.10.** Stolons grow above ground, continuously producing new plants. (Drawing by R. Castro.)

mother plant, can grow independently as new plant individuals. Stolons, on the other hand, are green, carry out photosynthesis, and creep above the ground in all directions, continuously forming new nodes. Mature nodes initiate new roots that anchor them to the soil when adequate moisture is present.

As a direct result of their growth habit, stoloniferous species such as St Augustine grass and centipede grass are easy to control along flower-beds, kerbs, etc. Periodic trimming of stolons prevents their encroachment into planters and similar areas. In contrast, rhizomatous species such as cynodon and zoysia grass can grow under kerbs or other shallow barriers. Planters or flowerbeds surrounded by these turfs are quite difficult to maintain. Appropriate growth habit is therefore a deciding factor in selection of turf-grass species in landscape areas. Rhizomatous species or species that produce both rhizomes and stolons are superior in areas subjected to frequent damage. Athletic fields should incorporate these types of turf, because otherwise bare spots are very slow to repair themselves. Stoloniferous species may require the overseeding or transplanting of large sections of turf (sod) from other areas.

## Roots

Roots also play an essential role in the life of the turf-grasses. Purely mechanically, roots anchor the plant in the soil. In addition, roots absorb water and nutrients from the soil and transport them to the stem; they store carbohydrates for later use and also send hormonal signals to the above-ground portion of the plant. Most turf-grass species have plentiful roots. A small patch of turf-grass may include thousands of individual plants. Each plant has its own fibrous root system intertwined with hundreds of root systems from neighbouring plants. Together, all these roots form a dense mat that may be much thicker and stronger than the above-ground portion of the turf. This layer of living and vigorous organic matter is truly efficient in absorbing water and nutrients that move down from the surface. Roots act as an efficient organic filter, catching almost everything useful that passes nearby (Fig. 1.11).

Roots of turf-grasses are sensitive to environmental changes and management practices. Because roots do not carry out photosynthesis, they depend entirely on the above-ground parts of the plant. Any stress to the plant top results in stress to the roots. Excessively low mowing height, for example, reduces photosynthesis and deprives the roots of food. Starving roots grow much more slowly and may even die. A mowing height sufficient to sustain root growth in the dry season, when sunshine is abundant, may be too low during the rainy season, when increased cloud cover dramatically reduces photosynthesis. Failure to increase mowing height can result in substantial thinning out of the turf, as roots suffer starvation. When warm and cloudy weather is combined with short days, the problem worsens because photosynthesis is reduced while the high temperature increases respiration, and carbohydrate reserves can be depleted even further. This situation arises in the tropics during the rainy season and is often viewed as the biggest challenge to tropical turf managers.

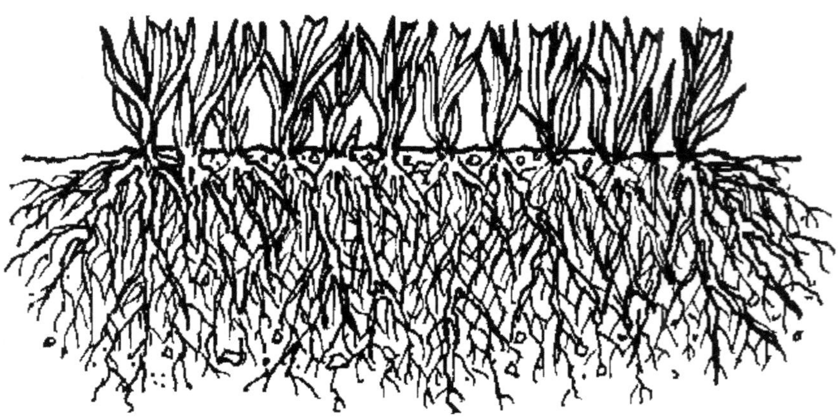

**Fig. 1.11.** The root system occupies a large volume of soil, efficiently absorbing water and nutrients. (Drawing by R. Castro.)

Overall the most crucial factor affecting root growth is mowing. Greater leaf area results in more photosynthesis, which in turn results in more roots and, especially important, more deep roots. Deep roots are able to extract water from deeper zones, reducing the chance of moisture stress and therefore improving the overall health of the turf and its resistance to other stresses. For this reason, areas where irrigation is not available should be mowed higher than irrigated ones, so that leaf area is not so severely reduced and root growth is less restricted.

Fertilization promotes rooting – adequately fertilized turf has a strong and healthy root system – but excessive nitrogen is detrimental to root growth. Because nitrogen promotes shoot growth, too much of it increases the shoot-to-root ratio. Rapidly growing shoots demand more energy, usually at the roots' expense. Avoiding excessive nitrogen fertilization is especially important when turf may be subject to other stresses, such as low light levels, drought and increased traffic.

Low light levels in tropical locations may greatly reduce rooting. Daylength near the Tropic of Cancer and Tropic of Capricorn ranges from 13.5 to 10.5 hours. Closer to the equator, the range is even narrower. Consider the performance of the same turf under temperate and tropical conditions. At moderate temperatures under 15 hours of sunlight, it may produce much more carbohydrate than it consumes. In the tropics, however, where the shorter days reduce photosynthesis but high temperatures maintain high respiration, less carbohydrate will be available to support root growth. The result is that carbohydrate reserves are often inadequate for unrestricted growth, especially where a distinct rainy season results in almost complete cloud cover, and therefore substantially less light, for several months during the year.

## Identification

Identification of grass plants requires careful observation of inflorescence (flower or seed-head), leaves and stems. Botanists around the world have described thousands of plants, not just grasses, using flowers as the major guide, so the inflorescence is probably the single most important component botanists use to identify grass species. In tropical turf-grasses, however, the inflorescence is not always present, and many turf-grass hybrids produce no inflorescence at all. In addition, managed turf-grasses are mowed frequently, so seed-heads are often difficult to find. Turf-grass managers therefore cannot depend on the inflorescence and must observe and use other parts of the plant for identification.

Unfortunately, no single characteristic can be used to identify individual turf-grasses with reasonable certainty. The process is usually complex. Numerous characteristics must be examined, and the information must be fitted together like the pieces of a puzzle. Only if all of the pieces fit is the identification convincing. Distinguishing between varieties within the same species can be very laborious, even when seed-heads are available. Sometimes only a DNA test can provide conclusive results.

Besides seed-heads, structures used for grass identification include the *sheath*, *collar*, *ligule*, *auricles*, *vernation* and shape of the *leaf tip*.

The sheath can be described as the lower part of the leaf that is attached to the node. It may be split along its full length from the collar to the node, split partway and closed for the rest of its length, or closed for its entire length (Fig. 1.12). The sheaths of most warm-season grasses are completely split, so the sheath is often not a very helpful characteristic for identification. An exception is carpet grass (*Axonopus compressus* (SW.) P. Beauv.), which has a flattened sheath that distinguishes it from most other warm-season species.

The collar is located at the junction between the leaf blade and the sheath. In many species the collar is absent or microscopic in size. In others it is quite visible and has distinctive characteristics like shape or colour. Either absence or presence of a collar can provide information useful in the identification process (Fig. 1.13).

The ligule grows from the collar area on the inner side of the leaf. It may be absent, take the form of a thin membrane, a fringe of hairs or a combination of the two. Most tropical turf-grasses have ligules, but sometimes they are too small to be seen without a magnifying glass. A ligule, if found, can be critical to identification (Figs 1.14 and 1.15).

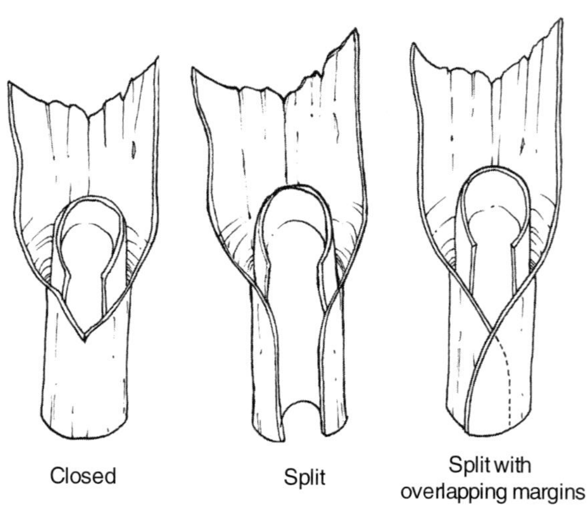


**Fig. 1.12.** Three basic types of sheaths. (Drawing by R. Castro.)

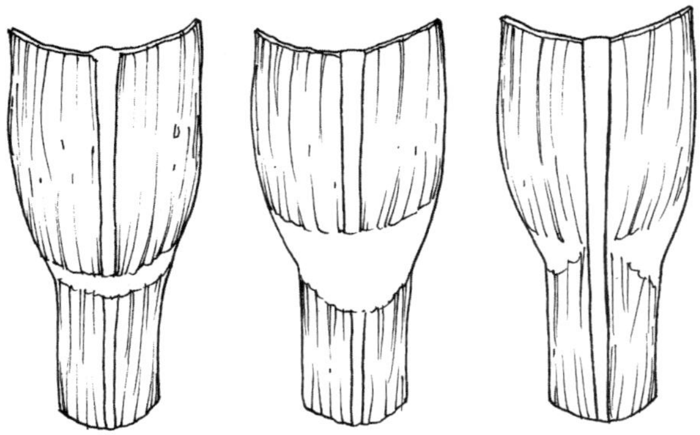


**Fig. 1.13.** Both size and shape of collar can help identify species. (Drawing by R. Castro.)

Auricles are appendages that sometimes grow from the edge of the collar and may wrap around the stem. Cool-season species have auricles, but tropical turf-grasses mostly do not (Fig. 1.16).

Vernation, the way new blades emerge from the sheath, is an important identification tool. In some species, leaf blades emerge rolled; in others they emerge folded (Fig. 1.17). Only one species, Bahia grass (*Paspalum notatum* Flüggé) often has both rolled and folded leaves on the same plant. This characteristic is very helpful in identification of this particular species. Without examination of vernation, distinguishing

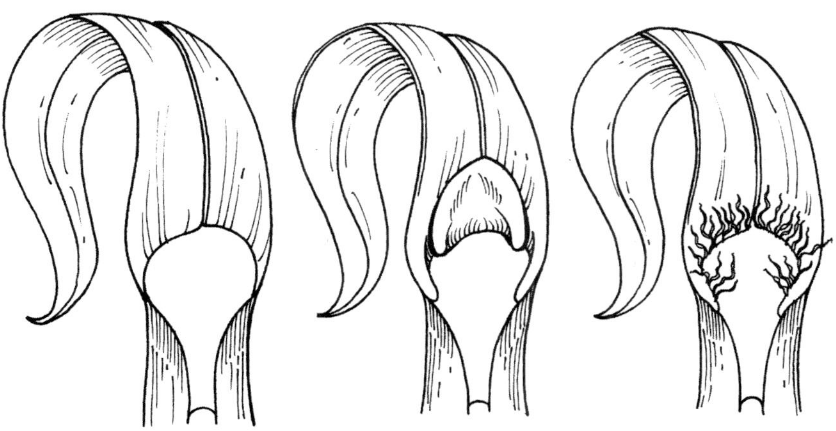


**Fig. 1.14.** Presence and type of ligule are important factors in species identification. (Drawing by R. Castro.)

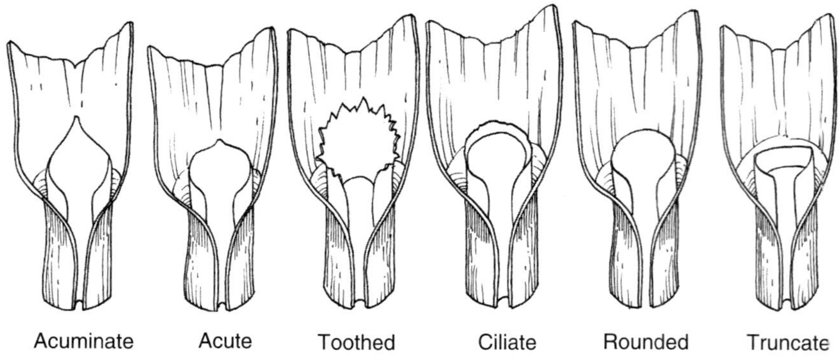


**Fig. 1.15.** Shape of ligule distinguishes many species. (Drawing by R. Castro.)

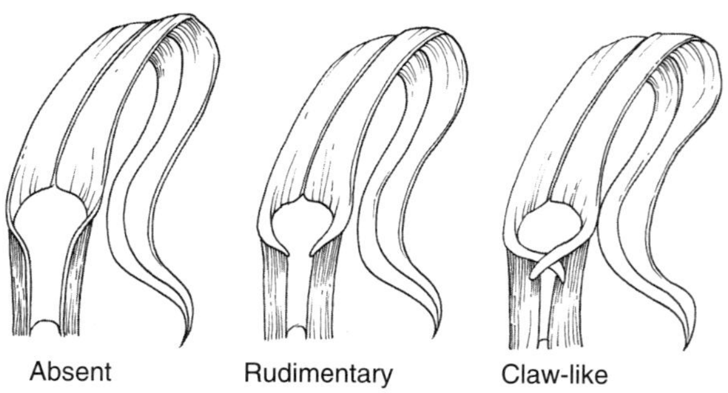


**Fig. 1.16.** Presence and type of auricles is most useful for identifying cool-season species. (Drawing by R. Castro.)

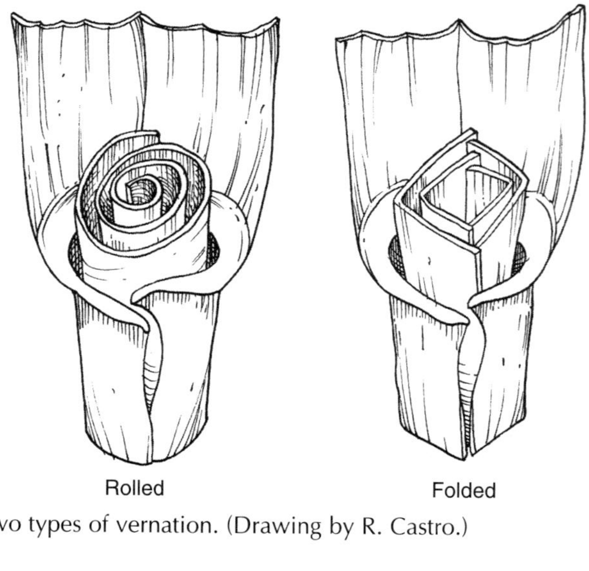


**Fig. 1.17.** Two types of vernation. (Drawing by R. Castro.)

cynodon (*Cynodon dactylon* (L.) Pers.) from some varieties of zoysia grass (*Zoysia* ssp.) could be difficult, but because cynodon has folded vernation and zoysia grass has rolled, identification is quite easy.

The leaf tip can sometimes play a role in species identification as well. St Augustine grass (*Stenotaphrum secundatum* (Walter) Kuntze), for example, has a characteristic rounded blade tip, which almost always provides positive identification of this species.

# 2 Turf-grass Species

All grasses belong to a single family of plants, the *Poaceae*. This family is divided into six subfamilies, which incorporate 25 tribes, 600 genera and 7500 species. Only 30–40 species are used as turf-grasses. All turf-grass species are within the three subfamilies: *Festucoideae*, *Panicoideae* and *Eragrostoideae*. Turf-grasses in the *Festucoideae* are usually adapted to cool climates. Those in the *Panicoideae* and *Eragrostoideae* are adapted to warm climates. This volume will address management of warm-season grasses only, but because some tropical locations in the mountains or other high elevations use cool-season species, a short description of major cool-season grasses appears at the end of this chapter. Cool-season and warm-season grasses require substantially different management practices, so the information provided in this volume may not be fully adequate for tropical locations with unusually cool climates.

## Warm-season Turf-grasses

### The genus *Cynodon (Eragrostoideae)*

*Cynodon dactylon* is known, in various regions around the world, by at least 20 different common names. The most popular are Bermuda grass, Bahama grass, devil's grass, couch grass, Indian doob, dog-tooth grass and wire grass. In this volume, the common name 'common cynodon' or just 'cynodon' will be used for *C. dactylon* (Fig. 2.1).

 

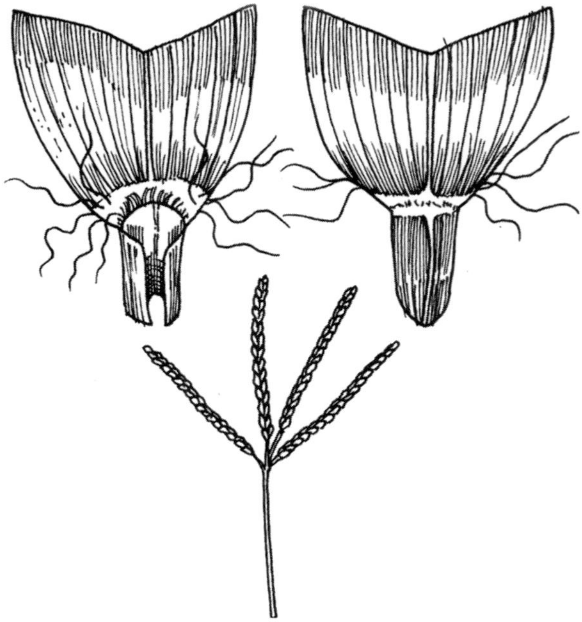

**Fig. 2.1.** Cynodon (*Cynodon dactylon*). Leaves folded in the bud; ligule a fringe of hairs; collar narrow, covered with long hairs; auricles absent; sheaths strongly compressed; blades short, 2–3 mm wide, rough along the edges, sharply pointed; seed-head 3–5 slender spikes, originating at the top of the main stem; spikelets oval, in two rows, each containing a single seed. (Drawing by R. Castro.)

Cynodon, which originated in Africa, besides being a very rich pasture plant, is the most widely distributed turf-grass around the world. It can be found in almost every region with a mild climate. In the tropical climate, common cynodon grows fast and continuously, usually staying dense and medium to dark green year-round. Where the soil temperature drops to 10°C for a long period, cynodon loses its chlorophyll and turns yellow or light brown and remains dormant until the soil temperature rises again and remains above 10°C. It is drought-tolerant and has a good tolerance to wear. It is well adapted to sunny conditions and has a medium-coarse texture with a greyish green colour. Common cynodon establishes a deep root system and produces long and lively rhizomes and stolons. The presence of vigorous rhizomes makes it a troublesome weed in flowerbeds that are adjacent to the turf. Cynodon rhizomes can pass under cement kerbs or other barriers and then emerge on the other side. The species is relatively resistant to numerous herbicides and other chemicals and to many adverse environmental conditions. Overall it is a very tough turf-grass. The only stress it does not tolerate well is low light intensity. In the humid tropics, especially during rainy periods lasting for several months, the light intensity is frequently reduced by heavy cloud coverage. Under these conditions cynodon thins out and to some extent loses its competitive advantage over weeds and other grasses. It also performs quite poorly under tree shade.

The genus *Cynodon* includes eight species besides *C. dactylon.* Several of them and their hybrids are used as turf-grasses and range in importance from major to marginal.

The *Cynodon dactylon* × *Cynodon transvaalensis* hybrid originated in the USA and since the 1970s has become the most important cynodon hybrid around the world. It is found predominantly on golf courses and on other recreational and sport turfs. It is continuously improved as new cultivars are released. Some well-known cultivars of this hybrid are available around the world and the most widely used are Tiflawn, Tifway, Tifgreen, Tifdwarf and TifEagle.

*C. transvaalensis* Burtt Davey, a very fine-textured turf species, also called African cynodon, has the softest leaf blades of all cynodon species. It is the parent, with *C. dactylon,* of many new fine-textured hybrids.

*Cynodon magennisii* Hurcombe, a natural hybrid of *C. dactylon* and *C. transvaalensis,* is very fine textured and is used for golf course greens in many parts of the world, especially Africa and Central and South America.

*Cynodon bradleyi* Stent, a fine-leaf turf species, is used in eastern and central Africa for golf course greens.

## The genus *Zoysia (Eragrostoideae)*

The genus *Zoysia,* which originated in Asia, includes 10 species, but only three are considered important turf-grasses: *Zoysia japonica* Steud., known also as Korean lawn grass or Japanese lawn grass; *Zoysia matrella* (L.) Merr., known as Manila grass; and *Zoysia tenuifolia* Willd. ex Thiele, also called Mascarene grass or Korean velvet grass.

*Z. japonica* is a creeping turf-grass that produces both stolons and rhizomes (Fig. 2.2). It grows well in a wide range of soils, in the full sun and under moderate shade. It tolerates heat and drought and is also the most cold tolerant of the zoysias. In the tropics, it grows well but much more slowly than cynodon. A smooth cynodon lawn can be established in 2–3 months, but a lawn of zoysia grass may take 8–10 months. The most obvious characteristic of this species is the stiffness of the leaf blades, which is caused by their high silica content. This morphological characteristic, easily recognizable to the touch, makes zoysia resistant to physical injury like that originating from foot or vehicular traffic. Zoysia's

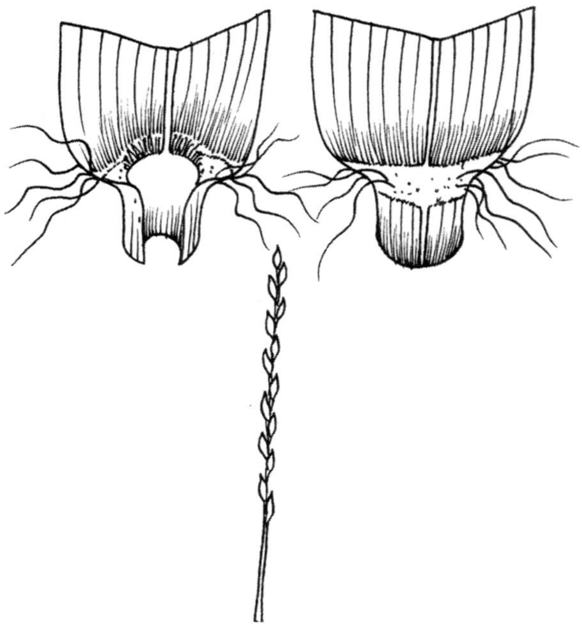

**Fig. 2.2.** Zoysia (*Zoysia japonica*). Leaves: rolled in the bud; ligule a fringe of hairs; collar medium, covered with long hairs; auricles absent; sheaths absent; blades short, 1–3 mm wide, sharply pointed; seed-head one weak spike with spikelets alternate on two sides; spikelets blunt at the base, round and tapering to a point. (Drawing by R. Castro.)

resistance to injury would make it a primary choice for sport fields that receive lots of concentrated traffic, like football fields and grass tennis courts, but its slow growth makes it slow to recover once it has been injured. In the tropical zone, therefore, *Z. japonica* is often chosen for golf courses, parks and playgrounds, and only sometimes for athletic fields. Homeowners do not consider zoysia the best choice for residential lawns, either. Because of its stiffness, it is not easy to mow, and mowing zoysia with dull blades, as often happens on home lawns, rips the turf surface or at best yields an uneven cut. An uneven, shredded cut not only looks ugly but also creates a perfect gateway for fungal infections and future health problems. For the best appearance, zoysia should be mowed regularly with a very sharp mower, preferably a reel mower (see Chapter 6, this volume). Overall, if properly managed, zoysia grass is relatively resistant to diseases. In the humid tropics it often suffers from insect damage. Sod web-worm is its worst pest and may infest this species several times a year. If overlooked during the time of massive invasion and not treated with chemicals, the sod web-worm can turn an entire lawn brown within a few days.

*Z. japonica* produces seeds, but its germination rate is quite low. In the tropics, seedlings take about 3–4 weeks to emerge from the soil, and in the meantime weeds may come to dominate the entire lawn area. Zoysia lawns are usually established vegetatively with sod, sprigs or plugs rather than by seeding (see Chapter 4, this volume). Besides sodding, sprigging is the fastest way to establish this species but nevertheless usually takes several months.

*Z. matrella* is similar to *Z. japonica*, except that it has a finer leaf texture and is less cold tolerant. It is found mainly in tropical and subtropical climates. *Z. matrella* does not produce many viable seeds and generally must be propagated from sprigs or plugs. It is only occasionally planted as residential lawn grass. Its overall importance is rather minor.

*Z. tenuifolia* is the finest textured of all zoysias. It does not produce viable seeds and must be propagated vegetatively. It can be found only in tropical and subtropical climates because of its very low tolerance to cold. Establishment of this zoysia is extremely slow. It will produce a thick thatch when left unmowed. This species is often used for residential lawns in the humid tropics because of its pleasant appearance and outstanding resistance to weeds. Unmowed turf areas of *Z. tenuifolia* have an unusual, non-uniform appearance, more like a ground cover than a lawn (Fig. 2.3). When mowed, it forms a pleasant lawn that is extremely resistant to traffic.

One interspecific hybrid of *Z. japonica* and *Z. tenuifolia* is worth mentioning. Emerald, which originated in the USA, forms a very dense

**Fig. 2.3.** *Zoysia tenuifolia* lawn, unmowed for many years.

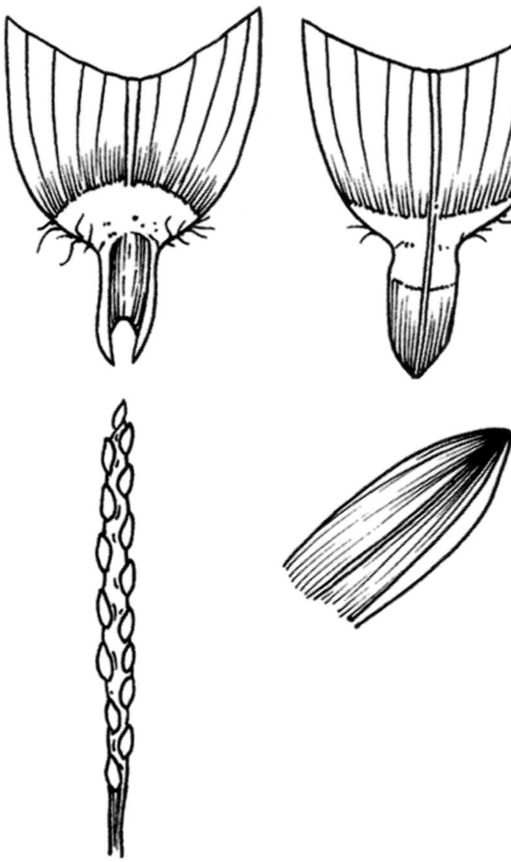

**Fig. 2.4.** St Augustine grass (*Stenotaphrum secundatum*). Leaves folded in the bud; ligule a fringe of short hairs; collar broad; auricles absent; sheaths compressed, flattened, with prominent mid-vein, sometimes sparsely hairy along the edges; blades short, smooth, 6–8 mm wide with 'boat-shaped' tips; seed-head a thick spike with few spikelets embedded along the sides. (Drawing by R. Castro.)

dark-green turf that performs very well in tropical climates. Availability of planting material outside the USA is limited.

### The genus *Stenotaphrum (Panicoideae)*

The single species in this genus is used as a turf-grass, *Stenotaphrum secundatum* (Walter) Kuntze, is most often called St Augustine grass, but buffalo grass is also a quite common name, especially in Australia (Fig. 2.4). St Augustine grass is a turf-grass widely adapted to the warm, humid regions of the world. It has a very coarse leaf texture and produces stolons but not rhizomes. Its large and fast-spreading stolons are easy to cut around flower beds and shrubs, either by hand or with a piece of turf equipment called an edger. This characteristic makes St Augustine grass a desirable choice for many homeowners and landscapers. It can be successfully grown in a wide variety of soils, with the exception of very alkaline soils, where it may require some supplemental iron fertilization. Under proper maintenance and a mowing height of 5–7 cm, it produces a dark-green, dense turf with exceptional shade tolerance. The shade tolerance of St Augustine grass is superior to that of all other tropical turf-grasses. St Augustine grass has poor tolerance for heavy traffic and cold temperatures and is quite often infested by insects, mostly by leaf-chewing caterpillars. It may also suffer from several fungal diseases as well as one viral disease, called St Augustine grass decline or SAD. Weed control in St Augustine grass is somewhat challenging, especially in the tropics. Few herbicides effectively control weeds without injuring the turf, but advances in biotechnological research are expected to solve weed-control issues in the future. St Augustine grass must usually be established by vegetative propagation in the form of sod, sprigs or plugs because its seeds are usually not viable. Numerous varieties of this species are available around the world.

### The genus *Eremochloa (Panicoideae)*

*Eremochloa ophiuroides* (Munro) Hack., called centipede grass or Chinese lawn grass, is very popular for use on residential lawns (Fig. 2.5). Subtropical and tropical climates are ideal for its growth. Centipede grass is a creeping grass of medium-coarse texture that produces above-ground stolons. It grows well in full sun and modest shade, but it does not tolerate heavy shade under trees. It is also not recommended for driveways or other areas subject to traffic.

The popularity of centipede grass in the tropics is related to its tolerance to low soil fertility. Under low fertility, centipede grass looks very pleasant and grows more slowly, so it needs less mowing than any of the other lawn grasses. If necessary, for example in the case of injury, its growth rate can be increased substantially by the application of fertilizer and unrestricted moisture. Then its growth can be slowed again by reversion to low-fertility management. Centipede grass ordinarily has a light green colour. High fertilization, especially with nitrogen, darkens the colour but results in numerous problems ranging from low stress tolerance to reduced resistance to weeds, insects and especially fungal diseases. When clippings are removed after mowing, centipede grass may

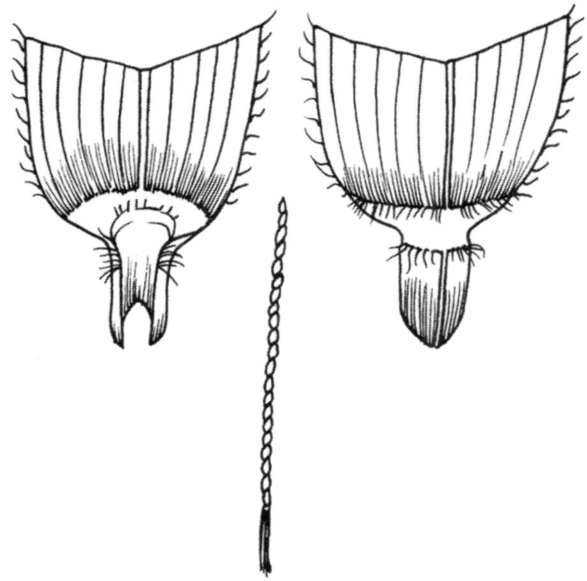

**Fig. 2.5.** Centipede grass (*Eremochloa ophiuroides*). Leaves folded in the bud; ligule a short membrane with short hairs across the top; collar broad, much constricted; auricles absent; sheaths compressed, flattened, hairs at the edges near the ligule, a prominent mid-vein; blades compressed or flattened, short, less than 4–6 mm wide, strong mid-vein, sparsely hairy along the edges; seed-head one slender spike at the top of the main stem; spikelets broad at the base tapering to a rounded tip, a single seed. (Drawing by R. Castro.)

require some fertilizer. If clippings are left on the ground, it usually does not need any fertilization at all. The most popular centipede grass is a common type that can be established from seeds. As with any other tropical turf-grass, centipede grass can also be established vegetatively from sprigs, plugs or sod.

## The genus *Paspalum (Panicoideae)*

*Paspalum vaginatum* Sw., known also as seashore paspalum or saltwater couch, is native to tropical and subtropical regions of North and South America, South Africa and Australia (Fig. 2.6). It produces few seeds, mostly not viable, and spreads by rhizomes and stolons. The stolons and leaves of seashore paspalum are slightly coarser than those of cynodon, but in fact these two species resemble one another. An important characteristic distinguishing paspalum from cynodon is the split shape of its

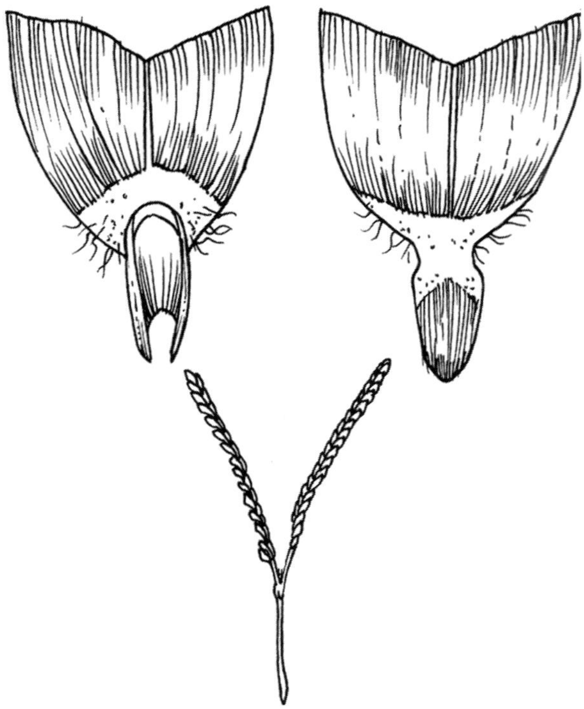

**Fig. 2.6.** Seashore paspalum (*Paspalum vaginatum*). Leaves folded in the bud; ligule a short membrane; collar constricted, broad, sparse long hairs at the edges; auricles absent; sheaths compressed, sparsely hairy; blades flat or folded, 4–6 mm wide, sparsely hairy, tapering to points; seed-head two slender spikes originating at the top of the main stem; spikelets blunt at the base, rounded at the centre, and tapering to blunt tips, each with a single seed. (Drawing by R. Castro.)

seed head. Besides *P. vaginatum*, only Bahia grass (*Paspalum notatum*) has a forking seed head, but Bahia grass is easily recognized by its substantially coarser leaves and much more open, transparent turf cover. In some parts of the world, such as South America and Asia, paspalum is used on golf course greens and tees. When mowed systematically, often twice a day, at heights of 3–4 mm, paspalum produces a very dense, excellent putting surface, sometimes better than those of hybrids of cynodon. One of the outstanding characteristics of paspalum is its tolerance to saline soils and saltwater spray. Along the ocean coast it also tolerates regular flooding with seawater. Some cultivars may carry on under saltwater irrigation. Paspalum needs less fertilizer than cynodon. Seashore paspalum can grow under conditions of low nutrient availability and severe nutrient imbalances. A deep root system makes it relatively drought resistant. It also tolerates wide pH ranges and low oxygen conditions. In the tropical climate, chemical weed control in paspalum is somewhat problematic. Post-emergence herbicides are often toxic to this species. Substituting saltwater applications for herbicide applications controls numerous weeds quite well.

*P. notatum* Flüggé, commonly called Bahia grass but also grama dulce or forquinha, is a low-maintenance turf-grass for poor soils (Fig. 2.7). It originated in South America but performs satisfactorily in Central America, Africa and Asia. It can be established from seeds or propagated vegetatively, mostly from sprigs or plugs. Bahia grass develops a sizeable root system, which makes it one of the most drought-tolerant turf-grasses. It does well in infertile soils, especially sandy soils, and does

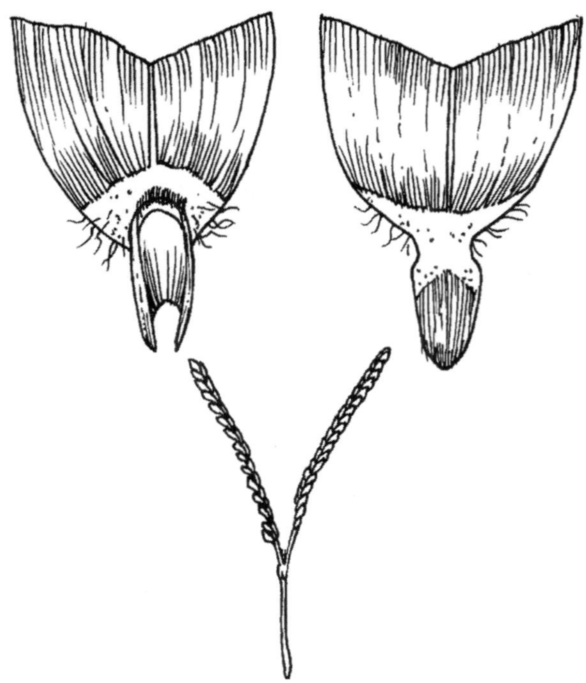

**Fig. 2.7.** Bahia grass (*Paspalum notatum*). Leaves folded in the bud; ligule a short membrane; collar constricted, broad, sparse long hairs at the edges; auricles absent; sheaths compressed, sparsely hairy; blades flat or folded, 4–6 mm wide, sparsely hairy, tapering to points; seed-head usually two slender spikes originating at the top of the main stem; spikelets blunt at the base, rounded at the centre, and tapering to blunt tips, each with a single seed. (Drawing by R. Castro.)

not require much, if any, fertilizer. Despite low maintenance requirements, Bahia grass must be mowed on a regular schedule, otherwise it produces tall and quite unsightly seed-heads. Unlike *P. vaginatum*, Bahia grass does not perform well in high-pH soils and does not have good tolerance to shade, traffic or salinity. Bahia grass has an open growth habit; that is, its stems grow more vertically than horizontally and often permit easier encroachment by weeds into open-space areas. Like *P. vaginatum*, Bahia grass has quite a low tolerance for many herbicides, making chemical weed control difficult.

## The genus *Axonopus (Panicoideae)*

*Axonopus affinis* Chase, known as carpet grass, and *Axonopus compressus* (Sw.) Beauv., known as tropical carpet grass, are coarse-textured, low-growing, light-green turf-grasses that spread by stolons (Fig. 2.8). Both species can be established from seeds or vegetative propagation. They are used mainly in low-quality turfs, along roadsides, on slopes, etc. Carpet grass tolerates low fertility and very acidic soils, requires little maintenance, and when established, holds on to soil on highly eroded slopes. Its importance is increasing especially in coastal areas with sandy soil.

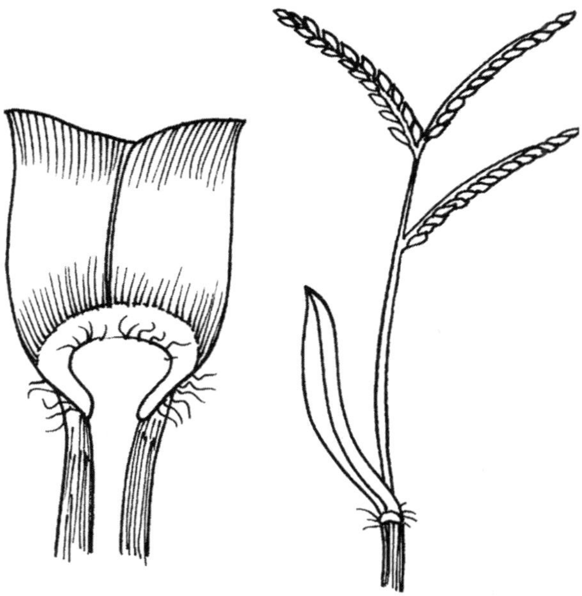

**Fig. 2.8.** Carpet grass (*Axonopus affinis*). Leaves folded; ligule a fringe of hairs fused at the base; collar narrow, continuous, sometimes with hairs; auricles absent; blades short, 4–8 mm wide, with short hairs near tips; seed-head three spikes; spikelets widely spaced in two rows. (Drawing by R. Castro.)

### The genus *Distichlis (Eragrostoideae)*

*Distichlis spicata* (L.) Greene, known as salt grass, alkali grass or spike grass, is a low-growing species that looks similar to low-quality cynodon. It can be used on all types of lawns, is excellent in parks and is often used in restoration of salt-marshes. Salt-grass seeds are mostly unavailable, and reproduction is primarily from rhizomes. Leaves are hairy and sharply pointed and have stiff blades. The sharp-pointed stiff rhizomes of this species can effectively push through heavy clay soils. Salt glands on the leaves extrude salt, allowing the plants to use salty water. The extensive root system of salt grass can hold together eroded soil particles, so it is a good erosion-control plant. It can occupy extremely salty and alkaline soils that are poorly drained and have a high water table. It can survive long-lasting soil saturation or flooding.

### The genus *Pennisetum (Panicoideae)*

*Pennisetum clandestinum* Hochst. ex Chiov., known mostly as kikuyu grass, is a coarse-textured, light-green grass that spreads by both rhizomes and stolons. It requires high fertility and moist soil that drains well. In the tropics it can be found mostly in mountainous regions of Africa and Central America and also in Thailand. When established in the hot, humid tropics it becomes severely diseased and cannot persist. Its importance is very minor.

## Criteria for Selecting Turf-grass Species

Correct selection of a turf-grass depends upon many factors. The most critical are type of use (lawn, golf course, sports field), location (sunny, shady, fertile soil, poor soil) and desired appearance. Because each turf-grass species has its good and bad characteristics, one should consider the strengths and weaknesses of each species for each particular situation.

The comments below are followed by tables ranking common turf-grass species according to their important characteristics or their requirements. Please note, however, that these rankings are not 'evenly spaced'. That is, two adjacent species in the ranking may differ a great deal or only slightly. For example, the recuperative capacities of first-ranked cynodon and second-ranked seashore paspalum are almost the same, whereas those of sixth-ranked centipede grass and seventh-ranked zoysia are very different. The exact position of a particular turf-grass in the rankings may depend somewhat on geographic location, but the general position (high, low or medium) is not likely to change and can be used as a general selection guide.

## Level of maintenance required

Because turf-grass species, even varieties, differ in required level of maintenance (Table 2.1), the grass selected must be appropriate for the level of maintenance that can be committed. Overall a much better turf will result from selection of a grass species that is less beautiful but can be maintained adequately than from selection of a more beautiful species whose maintenance needs cannot be met. Levels of maintenance are almost always directly related to cost and time. High-maintenance turf has the highest cost of adequate maintenance and requires the most maintenance time. Therefore, a realistic evaluation is required of the factors affecting maintenance: the cost of labour, availability and cost of equipment, availability and cost of irrigation water, personal preferences related to usage of chemicals etc. For example, a hybrid cynodon will form dense, carpet like turf but will not grow well without supplemental irrigation and will require frequent mowing and high fertilization. On the other hand, common cynodon may grow well with less irrigation, and will require less mowing and fertilization, but will not form as dense a turf.

## Establishment vigour

In the tropical climate establishment vigour is of major importance. Quick establishment is desired mainly because of weed pressure after seeding or vegetative planting. Highly competitive weeds may invade grass that establishes itself slowly. Some species establish themselves faster than

**Table 2.1.** Rankings of popular tropical turf-grasses, from highest at the top to lowest at the bottom, by level of maintenance required, establishment vigour, leaf texture (leaf width) and typical mowing height. Intervals between successive rankings are not constant.

| Maintenance required | Establishment vigour | Leaf width | Mowing height |
|---|---|---|---|
| Hybrid cynodon | Cynodon | St Augustine grass | St Augustine grass |
| Seashore paspalum | Seashore paspalum | Carpet grass | Bahia grass |
| St Augustine grass | St Augustine grass | Bahia grass | Carpet grass |
| Zoysia grass | Bahia grass | Centipede grass | Centipede grass |
| Common cynodon | Centipede grass | Seashore paspalum | Seashore paspalum |
| Carpet grass | Carpet grass | Zoysia grass | Zoysia grass |
| Centipede grass | Zoysia grass | Cynodon | Cynodon |

others (Table 2.1), but the more establishment time can be shortened the better. Establishment vigour is also discussed in Chapter 4 (this volume).

## Leaf texture

Relative blade width is called texture. Turf scientists use the three basic terms, coarse (wide), medium and fine (narrow), to describe texture, sometimes combining them to subdivide the continuum, as in 'medium-fine', 'fine-medium', 'coarse-medium' and 'medium-coarse'. Most often, texture is chosen simply by visual preference. Sometimes, especially for sport turfs, the necessity for a certain texture is dictated by the sport itself. Golf, tennis, cricket, hockey and numerous other games can only be played on turfs having particular texture. Table 2.1 shows common species ranked by texture.

## Mowing height

Each turf-grass species yields the best-quality turf at a particular mowing height (see Chapter 6, this volume). Some, such as hybrid cynodon, are mowed at 4–5 mm, whereas others, such as St Augustine grass, are mowed at 80–100 mm. Mowing too high usually reduces visual quality, but mowing too low may damage the turf. Mowing the turf below the recommended height reduces photosynthesis and therefore production of carbohydrates, which makes the turf weak and subject to invasion by weeds, insects and diseases. On the other hand, a moderate increase over the recommended mowing height is not harmful. For most turf areas it will somewhat reduce the optimal visual quality, but will result in an increased leaf surface that will promote more photosynthesis, a deeper root system, better drought tolerance and better overall turf health. On sports fields, mowing height is often imposed by the requirements of a game and cannot be increased.

Within the mowing-height ranking (Table 2.1) substantial differences may occur within a single species.

## Fertility requirements

Some turf-grasses have high fertility requirements, whereas others will perform satisfactorily with little or no fertilizer (Table 2.2). In fact, frequent fertilization of some of the less demanding species, such as Bahia grass or carpet grass, may result in serious weed, disease and insect problems. In contrast, the quite demanding cynodon may show

**Table 2.2.** Rankings of popular tropical turf-grasses, from highest at the top to lowest at the bottom, by fertility requirements, wear resistance, recuperative capacity and drought tolerance. Intervals between successive rankings are not constant.

| Fertility requirements | Wear resistance | Recuperative capacity | Drought tolerance |
|---|---|---|---|
| Cynodon | Zoysia grass | Cynodon | Cynodon |
| St Augustine grass | Cynodon | Seashore paspalum | Zoysia grass |
| Zoysia grass | Seashore paspalum | St Augustine grass | Seashore paspalum |
| Seashore paspalum | Bahia grass | Bahia grass | Bahia grass |
| Centipede grass | St Augustine grass | Carpet grass | St Augustine grass |
| Carpet grass | Carpet grass | Centipede grass | Centipede grass |
| Bahia grass | Centipede grass | Zoysia grass | Carpet grass |

symptoms of severe nitrogen deficiency within a few months without fertilization, especially in the rainy season.

## Wear resistance and recuperative capacity

The genetically determined ability of turf-grass to resist forces that crush leaves, stems and the crown of the plant is known as wear resistance or wear tolerance. It is an important characteristic of particular turf species, especially on highly trafficked areas such as athletic fields, playgrounds, and golf course greens and tees. Turf-grasses that are established in these locations are subject to damage from foot traffic, bouncing balls and light vehicular traffic. When an injury does occur, turf-grasses differ in their ability to recover, called recuperative capacity. Restoration of full vigour and pleasant appearance may take one species a few days and another several weeks. Zoysia grass has the highest resistance to injury (has high wear resistance), but if injured it takes the longest to recover (has low recuperative capacity) (see Table 2.2).

## Drought resistance

The most frequent stress endured by turf-grasses is drought, the lack of water. Drought tolerance describes the ability of a turf to hold up well over dry periods without irrigation or rainfall. Overall, tropical turf-grasses survive long-lasting drought well. Some species, such as cynodon, roll their leaves and stop growing but seldom turn brown. Others lose their green colour and remain yellow to brown for many weeks. They may look dead, but when water becomes available, they quickly regain their colour and healthy appearance. Table 2.2 shows common species ranked by drought tolerance.

## Shade tolerance

Another frequent cause of turf-grass decline is shade. Under heavy shade, turf becomes weak or simply dies; this effect is often visible under the trees or along the shaded sides of buildings. Cynodon does not tolerate shade and should never be selected for shaded areas. St Augustine grass tolerates shade well and is therefore usually preferred for home lawns and other significantly shaded areas. Table 2.3 shows common species ranked by shade tolerance.

## Traffic tolerance

Almost every turf area receives some traffic. Some turfs, such as home lawns, receive little; others, such as sport fields, receive much more. Distribution of traffic is often uneven, e.g. on golf courses, home lawns and sport turfs. Areas close to the entry usually receive most of the traffic, golf course greens receive more than fairways, and areas in front of the goals on a football field receive more than the areas close to the side-lines.

When a turf-grass is selected, the most heavily trafficked areas should be considered, rather than the areas under lighter traffic. The ranking based on traffic tolerance is the same as that for wear resistance (Table 2.2).

**Table 2.3.** Rankings of popular tropical turf-grasses, from highest at the top to lowest at the bottom, by shade tolerance, salt tolerance, acid tolerance, heat tolerance and cold tolerance. Intervals between successive rankings are not constant.

| Shade tolerance | Salt tolerance | Acid tolerance | Heat tolerance | Cold tolerance |
|---|---|---|---|---|
| St Augustine | Seashore paspalum | Carpet grass | Zoysia | Zoysia |
| Zoysia | Cynodon | Centipede | Cynodon | Cynodon |
| Centipede | Zoysia | Cynodon | Carpet grass | Seashore paspalum |
| Carpet grass | St Augustine | Seashore paspalum | Seashore paspalum | Bahia grass |
| Bahia grass | Bahia grass | Zoysia | Centipede | Centipede |
| Seashore paspalum | Carpet grass | St Augustine | St Augustine | Carpet grass |
| Cynodon | Centipede | Bahia grass | Bahia grass | St Augustine |

### Tolerance of salt stress

In the dry tropics, salts are brought to the soil surface mainly as a result of imbalance between evaporation and infiltration of rainwater. Other major sources of salt accumulation are saltwater flooding, saltwater intrusion and saltwater (ocean) spray.

Turf-grass species differ considerably in tolerance to salt. The most resistant varieties of seashore paspalum may tolerate long-term irrigation with ocean water (over 3% NaCl), whereas centipede grass may be severely injured by one saltwater spray event (Table 2.3).

### Tolerance of soil acidity

High soil acidity (low pH) can be corrected by application of lime (calcium hydroxide), but liming may not be practical along roadsides or on large turf areas subject to low maintenance. Therefore, species tolerating high soil acidity are frequently selected for these areas. Of the warm-season grasses, carpet grass and centipede grass tolerate low pH quite well. Bahia grass and St Augustine grass do not (Table 2.3).

### Heat tolerance

Tropical turf-grasses are more frequently stressed by extreme heat than by cold. Hot days followed by very cold nights sometimes occur in tropical desert areas, especially at higher elevations. Zoysia grass and cynodon tolerate both heat and cold well. St Augustine grass requires mild conditions (Table 2.3).

## Cool-season Turf-grasses

Turf-grasses having a temperature optimum of 15–22°C are referred as cool-season turf-grasses. These species are widely distributed in cool climates but can be also found in some tropical locations, for example at high elevations. The most important cool-season grass species are members of the four genera *Poa* (the blue grasses), *Festuca* (the fescues), *Agrostis* (the bent-grasses) and *Lolium* (the rye-grasses).

### The genus *Poa (Festucoideae)*

The highly diverse genus *Poa* comprises more than 500 annual and perennial species, including bunch-grasses (species that tend to grow in clumps), rhizomatous grasses and stoloniferous grasses. Only seven *Poa* species are used as turf-grasses, and only three are of significant importance.

*Poa pratensis* L. is the most widely used of the cool-season turf-grasses throughout the world. It can be found on lawns, golf courses, cemeteries, parks, school grounds, athletic fields and other areas where a dense grass cover is desired. *Poa pratensis* recovers from injury quickly and spreads very fast. It develops a dense turf, has excellent colour and is easy to mow. It tolerates low temperatures and many diseases. During periods of drought, it undergoes dormancy but easily survives and can quickly regain its colour when water becomes available. When mowed at the correct height, 50–70 mm, it competes very successfully with weeds.

Disadvantages of this species include its shallow root system, relatively high demand for water and poor tolerance of shade. Hundreds of *P. pratensis* cultivars have been developed around the world. Old, common types are well adapted to lower-maintenance conditions. New, improved cultivars are suitable for high-maintenance turfs.

*Poa trivialis* L. is used in cool, humid regions in wet, shaded areas along rivers and lakes and where conditions are too shady for *P. pratensis.* It does not tolerate traffic or drought well. *P. trivialis* is sometimes used for winter overseeding of dormant cynodon turf (see Turf-grass Communities, later in this chapter).

*Poa annua* L. (annual blue grass) has the widest geographic distribution among grasses. It is often considered a weed, but in some countries it has become so dominant, and the possibility of eradication so remote, that it is used as the primary turf species. It can be found nearly everywhere in the world except the humid tropics. Its extraordinary success is related to the timing of its germination. Unlike other weedy grasses, which germinate in the spring, *P. annua* germinates in the late summer and early autumn; it lives through the winter as a mature plant and then, after producing seeds in the spring, it dies. Annual blue grass is particularly well adapted to low mowing heights, and it is on closely mown turf that it generally becomes a problem.

Other *Poa* species, of minor importance as turf-grasses, include *P. supina* Schrad., *P. compressa* L., *P. bulbosa* L. and *P. alpina* L.

## The genus *Festuca (Festucoideae)*

The genus *Festuca* includes more than 360 species that differ widely in appearance. Less than ten species are used as turf, all in cool climates.

*Festuca arundinacea* Schreb., tall fescue, is a deep-rooted, cool-season perennial grass. It produces vigorous growth in spring and autumn, and its extensive root system helps it withstand drought conditions. The species is adapted to a wide range of soil and climatic conditions, but performs best where winter is rather mild. Its requirement for relatively high mowing limits its use to lawns, parks, golf course roughs and other areas mowed at 40 mm or more. Proper, frequent watering is important to its survival. Major insect problems include army-worms, cutworms and white grubs. It is fairly tolerant to most turf-grass diseases. Many *F. arundinacea* lawns become thin and unattractive after hot, dry summers, requiring application of new *F. arundinacea* seed every autumn.

*Festuca rubra L.* is best adapted to cold northern climates. It tolerates cold temperatures, but not heat or drought.

*Festuca rubra* ssp. *fallax* (Thuill.) Nyman is best adapted to shade, but will persist in full sun. It tends to tolerate summer stress a little better than does *F. rubra*.

*Festuca brevipila* Tracey is better adapted to drier areas than *F. rubra*, but tends to thin out in wet years. It is an excellent addition to shade mixtures for lawns, but will also grow well in full sun.

*Festuca ovina* L. is generally used in low-maintenance areas.

## The genus *Agrostis (Festucoideae)*

The genus *Agrostis* is composed of about 220 species, but only a few are suitable for use as turf. All species have fine leaf texture and are adapted to low mowing heights.

*Agrostis palustris* Huds. is used mainly on high-quality turfs maintained at low mowing heights. It is the primary cool-season grass for golf greens in the temperate regions of the world. It can be used also on fairways and tees if mowed a little higher. *A. palustris* is seldom used for lawns. It requires intensive care, specialized mowing equipment and a generally high level of maintenance. *A. palustris* has a relatively shallow, but very dense, root system.

*Agrostis capillaris* L. is a fine-textured, sod-forming turf-grass. Unlike *A. palustris* it is not well adapted to very low mowing heights, so it is better suited for golf course fairways than for greens or tees.

*Agrostis canina* L. is the finest textured of the *Agrostis* species. It forms more beautiful putting surfaces than any other warm- or cool-season turf-grass. It grows well along coastlines and other areas where the weather is mild, cool and moist.

*Agrostis* species of rather minor importance as turf-grasses include *Agrostis alba* L., *Agrostis castellana* Boiss. et Reuter and *Agrostis idahoensis* Nash.

## The genus *Lolium (Festucoideae)*

*Lolium* includes eight species. Their most important characteristic is their rapid seed germination and establishment.

*Lolium perenne* L., perennial rye-grass, is a bunch-grass known for its rapid germination and establishment. It is useful for quick repair of damaged areas on lawns, athletic fields and golf-courses. *L. perenne* is sometimes used in combination with *Poa pratensis* on lawns and athletic fields, especially under heavier traffic. In warmer regions, it is used for winter overseeding of dormant warm-season turf. *L. perenne* is best adapted as a permanent turf-grass where winters and summers are moderate and moisture is sufficient.

*Lolium multiflorum* Lam., known in many countries as Italian rye-grass, grows very rapidly. It can be used for lawns, but is usually reserved for specialized uses such as turf repair, quick temporary turf cover or overseeding.

### Other cool-season genera

Other, less important, cool-season turf-grasses include *Bromus inermis* Leyss. (smooth brome-grass), the genus *Phleum* (timothy grasses), *Dactylis glomerata* L. (orchard grass) and *Puccinellia distans* (L.) Parl. (reflexed salt-marsh grass). Almost certainly several other cool-season species are used as turf-grasses in various regions of the world, but most are of minor importance.

## Turf-grass Communities

An individual turf-grass plant lives in a *community* with many other plants, which may or may not be of the same genus, species or cultivar. A turf-grass community can be classified as either a *monostand* or a *polystand*. Monostands are composed of turf-grass plants belonging to only one species and cultivar. *Polystands* are composed of turf-grass plants of more than one cultivar and/or species. Polystand turf-grass communities can be further subdivided into *blends,* composed of two or more cultivars of a single species, and *mixtures,* composed of two or more different species.

Blends and mixtures are very common in cool-season turf communities. They increase genetic diversity; community resistance to diseases, insects, environmental stresses; etc. For example, if one of three varieties or species within a turf-grass community is damaged or killed by a disease or insect, the other two may be unaffected and may grow to take the place of the missing plants, preserving the turf's overall appearance. The most important condition for usage of mixtures and blends is compatibility of the component varieties or species. Mixed or blended turf-grasses should have similar leaf texture, growth habit, colour and shoot density, and fairly equal growth vigour. Unfortunately, warm-season (tropical) grasses rarely meet these criteria, so mixtures of warm-season turf-grass species are rarely used. When one species (or even variety) occurs within a monostand of another species, it usually disrupts the pleasant, consistent appearance of the turf area and becomes regarded as an undesirable weed. The only instances in which mixed turf-grass communities might be desirable in the tropics are those calling for *winter overseeding*. At high elevations, where winter temperatures drop below 10°C and turf-grasses go dormant for several months, cool-season species may be overseeded into warm-season species to provide a green turf while the warm-season species is brown and dormant. With the coming of spring, the cool-season species gradually thins out and the warm-season species takes the lead. To be effective in winter overseeding, the cool-season turf-grass species should provide the shortest possible transition time from a green warm-season turf to a green cool-season turf and a similar smooth transition back to the green, warm-season turf in the spring. Overseeding is usually expensive and reserved for high-quality turfs. Seeding rate is often 5–10 times higher than that used for regular turf establishment, and because virtually all plants of the cool-season grass die during the following summer, overseeding must be repeated every year.

# 3 Soil

Soils of the tropical regions of the world are extremely diverse. They range from the poor arid and semi-arid soils, which are low in organic matter, to the humid tropical soils, which are relatively high in organic matter and more fertile. Because high temperatures persist throughout the year, as does high humidity in some areas, microbiological and chemical activities in tropical soils are much faster than those of the temperate regions. Rainfall is often seasonal, sometimes excessive and, in some areas, abundant year-round. The information included in this chapter pertains to the general properties of soil that are typical for most tropical regions.

Soil is the material that covers the surface of the earth, whenever that surface is not bare rock. It can be formed by the forces of nature or be built up artificially. Soils used in crop production are usually natural, whereas those under turf-grass are often artificial, composed of material brought from elsewhere.

Soil consists of four major components: (i) minerals, (ii) air, (iii) water and (iv) humus (Fig. 3.1). Minerals are inorganic substances, originating from rock. Air and water fill the gaps between minerals. Humus is organic material originating from living sources.

All soil components influence each other and form the environment in which plants grow. If this environment is well balanced – i.e. if it contains adequate amounts of water, air and all necessary nutrients in the right proportions – the result is a fertile soil.

The basic functions of soil are to provide physical support for the turf-grass plants and to supply moisture, air and nutrients to the roots.

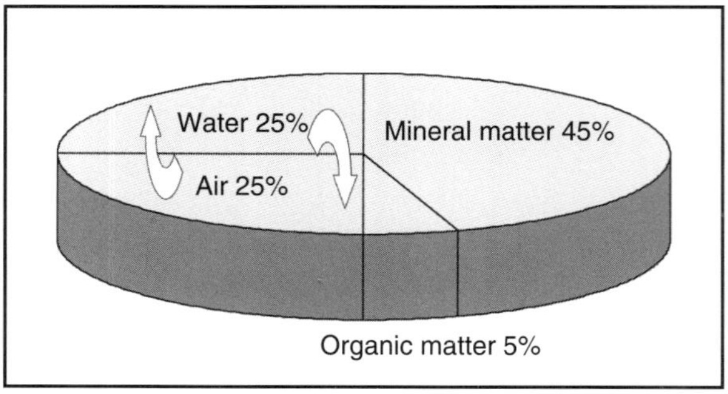


**Fig. 3.1.** Soil is composed of minerals, air, water and organic matter (humus).

Successful management of turf requires familiarity with the essential components of soil, their basic physical and chemical properties, their role in the soil and their influence on development of plants. Proper management of soil is essential in successful management of turf.

## Soil Texture

The term soil texture describes the size of the individual mineral particles and the proportion of each size in the soil The largest soil particles are called *sand*, intermediate particles are called *silt* and the smallest particles are called *clay*. According to classification adopted by most countries, sand consists of particles ranging from 1 mm (in some countries 2 mm) to 0.05 mm in diameter. Silt consists of particles ranging from 0.05 mm to 0.002 mm (in some countries 0.02 mm). Clay consists of particles smaller than 0.002 mm in diameter. Figure 3.2 shows the relative sizes of these types of soil particles. If the sand particle were the size of a basketball, a silt particle would be the size of tennis ball and a clay particle would be smaller than a grain of rice.

Soil types are based on texture. For example, sandy soils are those that contain a high proportion of sand particles, and clayey soils are those containing a high proportion of clay. Soils containing a high proportion of silt, however, are usually called loamy rather than silty. On the basis of the proportions of sand, silt and clay particles, distinct soil textural classes have been defined. Classification differs slightly in different countries, but patterns are similar. Soils containing primarily sand and silt are called loamy sands or sandy loams. Increasing amounts of clay could result in a sandy clay loam or sandy clay. Soil that consists of roughly the same percentages of sand, silt and clay is classified as a clay loam. The *textural triangle* shown in Fig. 3.3 presents 12 textural classes and is the most commonly used around the world.

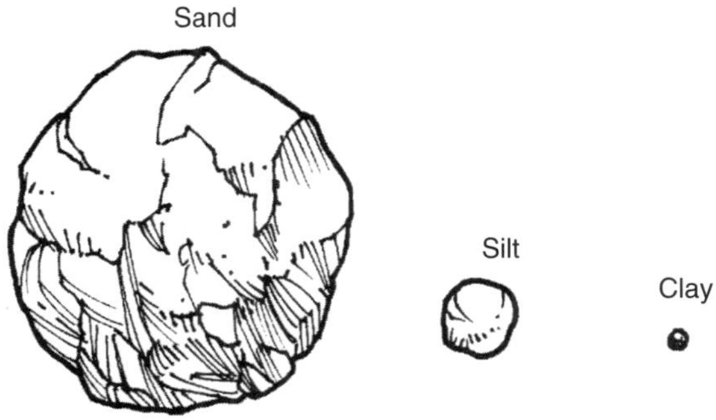


**Fig. 3.2.** Relative sizes of three major types of soil particles. (Drawing by R. Castro.)

## Soil Structure

Development of a stable soil structure often helps to ease the challenges resulting from imperfect soil texture. Soil structure is the way in which individual soil particles are bound together into larger units called *aggregates*. With some variation around the world, soils are usually classified as structural, moderately structural or weakly structural. Aggregates are generally described as blocky, granular, platy or prismatic (Fig. 3.4). Organic matter is the most important factor contributing to stable soil aggregation. Ideal soils are composed of approximately one-half solid material and one-half pore space. Organic matter helps to achieve this ideal composition. Organic matter in the soil undergoes quite complex biological and chemical transformations, glues soil particles together, helps to create small pores within each aggregate and larger pores between aggregates, and in this way improves aeration, moisture retention and air movement, and makes the entire soil more fertile. In addition, soil organic matter helps to supply available plant nutrients, decrease loss of soil to erosion, and improve soil's cation exchange capacity.

## Soil Porosity

Soil texture and structure directly affect the volume of pore space in the soil. For example, as shown in Fig. 3.5, clayey soil contains considerably more total pore space than a sandy or loamy soil. Because pore spaces are occupied by water and air, both needed by plants, having the highest possible pore volume might seem beneficial. In fact, it often is beneficial,

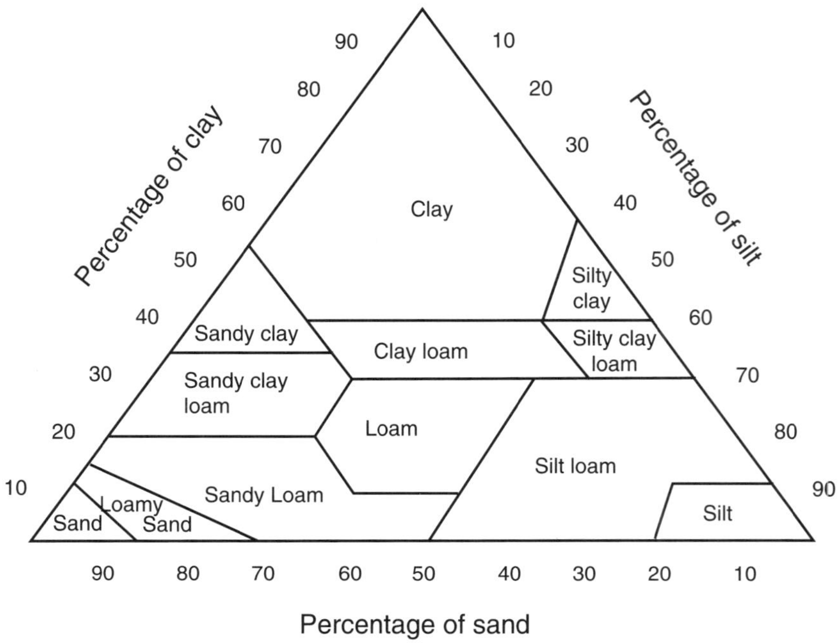


**Fig. 3.3.** Widely accepted textural triangle showing the names commonly assigned to soil textural classes according to their relative percentages of sand, silt and clay.

but not always. If the pores are too small, they will exert strong capillary forces and hold moisture very tightly – sometimes so tightly that roots are unable to draw it away. In this case, turf may suffer water stress even when porosity is high and the percentage of water in the soil is high. On

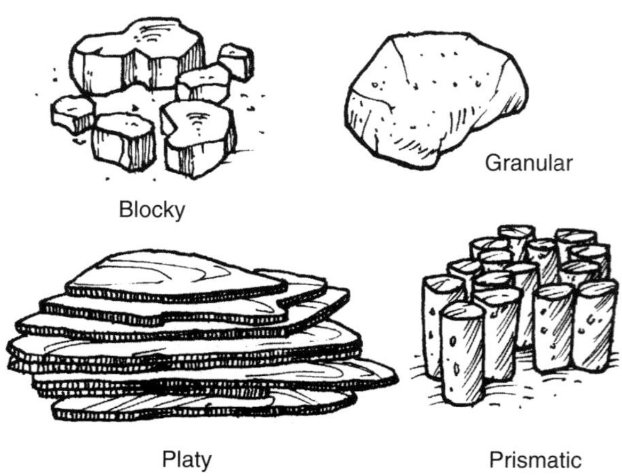


**Fig. 3.4.** Our basic shapes of soil aggregates. (Drawing by R. Castro.)

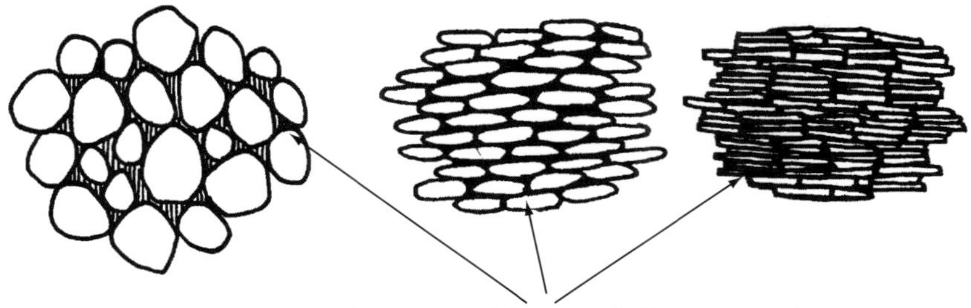


**Fig. 3.5.** Total porosity of clay is much larger than that of sand, but the pores may be too small to benefit plants. (Drawing by R. Castro.)

the other hand, soil with large pores and weak capillary forces may hold only a small percentage of water, allowing the rest to drain by gravity before it can be absorbed by roots (Fig. 3.6). Not only is this water lost, but it may carry dissolved nutrients away as well.

## Soil Cation Exchange Capacity

During physical and chemical weathering, soil minerals and organic matter break up into very small particles called colloids. These particles are generally negatively charged and can therefore attract and hold positively charged ions, called cations. Many plant nutrients, such as potassium ($K^+$), ammonium ($NH_4^+$), calcium ($Ca^{2+}$) and magnesium ($Mg^{2+}$) exist in soil solution as cations and are attracted by negatively charged colloids. The cations are held in what is called exchangeable form; i.e. they can be exchanged with other cations and become available

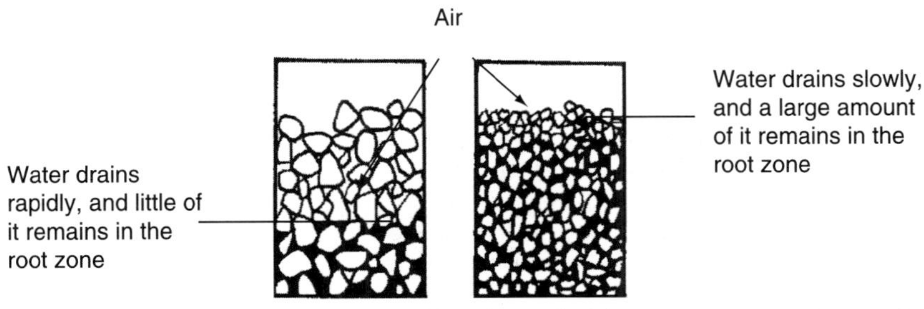


**Fig. 3.6.** In more porous soil, the capillary forces holding water are weaker, and it may drain away before roots can absorb it. (Drawing by R. Castro.)

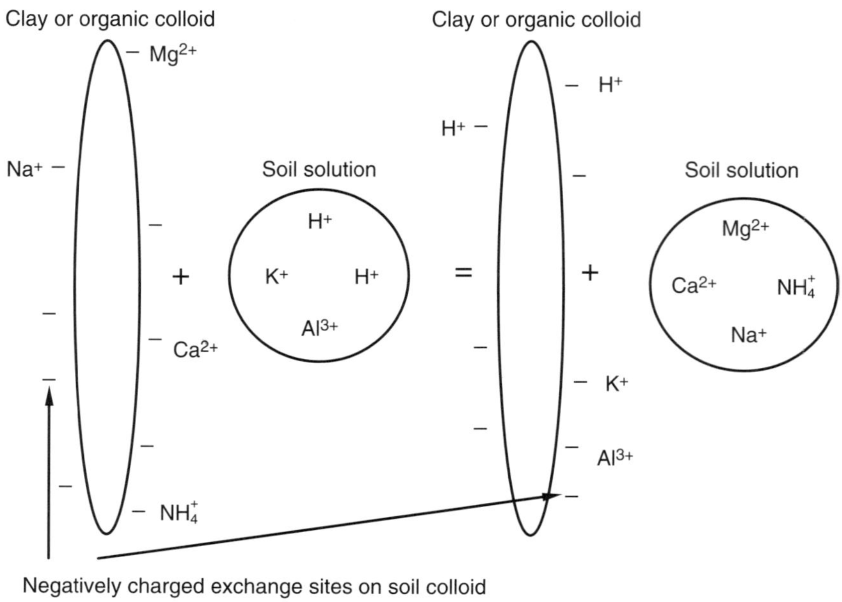


**Fig. 3.7.** Exchange of cations between soil colloids and the soil solution provides a continuous supply of nutrients available to plants.

to plants (Fig. 3.7). The ability of a soil to hold cations in this form is known as *cation exchange capacity* or CEC. It is one of the most important chemical properties of soil and is usually expressed in units called milliequivalents per 100 g of soil (meq/100 g). The CEC of a soil might be compared to the size of a car's fuel tank. The larger the fuel tank, the longer the car can travel before refuelling. Likewise, the larger the CEC, the more nutrients the soil can hold and the longer it can sustain plant growth without replenishment. In general, the more clay or organic-matter colloids a soil contains, the higher its CEC (Table 3.1). In practical terms, turf on low-CEC (sandy) soil may need fertilization every two to three weeks, whereas turf established on a high-CEC (clay loam) soil may need fertilization only every 2 to 3 months.

## Soil pH

Soil pH is another important chemical property. Mathematically, pH is defined as the negative logarithm of the hydrogen-ion concentration. In other words, the pH is a logarithmic expression of the concentration of

**Table 3.1.** Examples of cation exchange capacities (CEC) for different types of soils.

| Type of soil | CEC, meq/100 g of soil |
|---|---|
| Light-coloured sands | 1–5 |
| Dark-coloured sands | 5–20 |
| Light-coloured loams and silt loams | 10–15 |
| Dark-coloured loams and silt loams | 15–25 |
| Clay and clay loams, loams and silty clays | 20–50 |
| Organic soils | 50–100 |

hydrogen ions ($H^+$) in the soil solution (the water in the soil and all the substances dissolved in it). The pH scale ranges from 0 to 14, and pH is expressed without units of measure. A pH value of 7.0 is neutral; i.e. at pH 7.0, the concentrations of hydroxyl ions ($OH^-$) and hydrogen ions ($H^+$) are exactly equal. At pH values below 7.0, the concentration of hydrogen ions, written '$[H^+]$', is greater than that of hydroxyl ions, written '$[OH^-]$', and the soil is called acidic. At pH values above 7.0, $[OH^-]$ is greater than $[H^+]$, and the soil is called basic or alkaline. Most turf-grasses require a soil pH between 5.5 and 7.5.

## Soil Buffering Capacity

Another important CEC-related property is *soil buffering capacity*. Because buffering means 'resisting rapid change', soil buffering capacity tells how resistant the soil is to rapid change of pH. Buffering capacity can be compared to the diameter of a container. Raising the water level by, for example, 10 cm in a narrow container takes much less water than raising the level in a wide container by the same amount (Fig. 3.8). Likewise, raising the pH of a sandy, low-CEC soil into the acceptable range requires the addition of much less highly alkaline lime than does raising that of a loamy, high-CEC soil starting from the same low pH.

Knowing a soil's CEC and buffering capacity is important, because they are directly related to the amount of liming material required to adjust soil pH. Some soils in the tropics can be very acidic (have very low pH) and must be limed before turf-grass can be established. Only a laboratory test of CEC and buffering capacity can reveal how much lime should be applied. Laboratories usually present their results in the form of a liming recommendation. Table 3.2 provides approximate quantities of the most common lime source (calcium carbonate, $CaCO_3$)

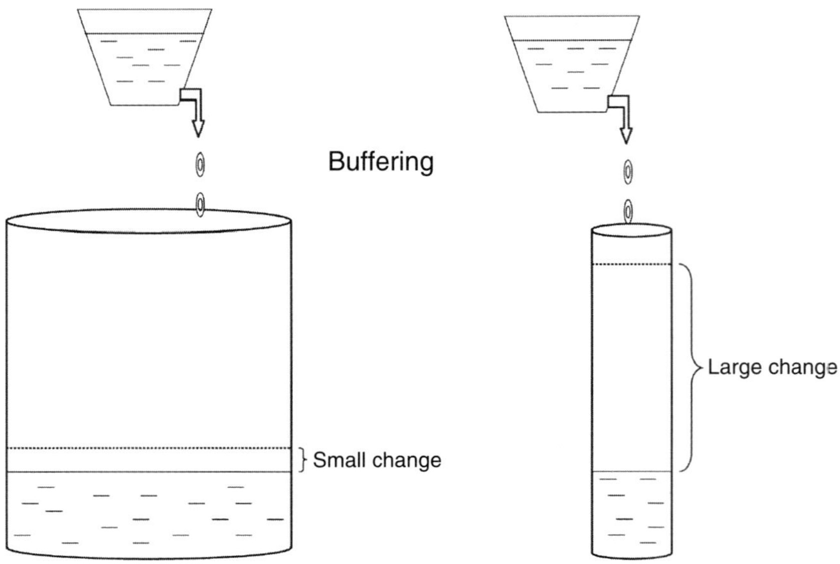


**Fig. 3.8.** When an equal volume of water is added to each tank, the level in the slim tank will rise significantly, but that in the wide one will rise only slightly. Similarly, addition of the same amount of lime may drastically change the pH of low-CEC soil and have only a small effect on the pH of high-CEC soil.

required to adjust the pH of sands, loams and clays to levels suitable for turf-grass.

## Effect of Soil pH on Nutrient Availability

Nutrient availability is greatly influenced by soil pH (Fig. 3.9). Correct pH is especially important for phosphorus availability, which is greatest in neutral soils. In acidic soils, phosphate ions bond with iron and aluminium to form insoluble iron and aluminium phosphates. At pH levels above 7.0, insoluble calcium and magnesium phosphates are formed. Phosphorus solubility, and hence availability to plants, is greatest at a soil pH between 6.0 and 7.5. Soil pH strongly affects availability of

**Table 3.2.** Approximate quantities of ground $CaCO_3$ (kg/100 $m^2$) required to raise pH to 6.5 in a 15-cm-thick root zone.

| Original soil pH | Sand and loamy sand | Loam | Clay loam and clay |
|---|---|---|---|
| 5.5 | 7–15 | 20–40 | 40–50 |
| 4.5 | 15–25 | 40–70 | 80–100 |

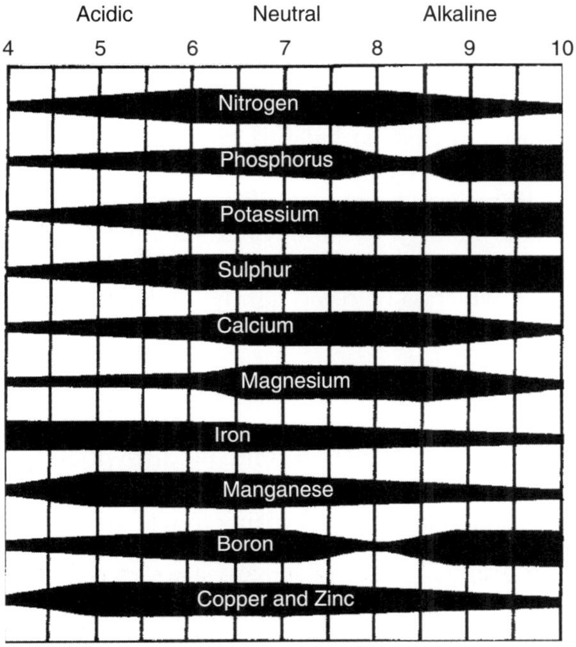


**Fig. 3.9.** Availability of essential nutrients depends strongly on soil pH (bar thickness corresponds to relative availability of the nutrient). (Drawing by R. Castro.)

other nutrients as well. In general, the availability of nitrogen, potassium, calcium and magnesium is high in neutral soil and decreases rapidly when pH drops below 6.0 or rises above 8.0. Iron, manganese, boron and several others are most available when soil pH is in the acidic range. As the pH of soil approaches the otherwise optimal 7.0, their availability decreases, and deficiencies can become a problem. Turf-grasses in the tropics often suffer from iron deficiency, especially on alkaline, coral soils along the coast or on sodic (sodium-rich) soils in arid climates. Sometimes only foliar applications of iron and other micronutrients is effective. Because the balance between soil pH and nutrient availability can be so fine, soil acidity should be tested periodically and maintained within the optimum range.

## The Soil Water

Most of the water in soil originates from natural precipitation, irrigation or (occasional) flooding and is held in the pore space between the solid particles of minerals and organic matter. After an intense rain or excessive irrigation, the soil is saturated; all the pores are filled up. With time, water is removed from the large pores by gravitational forces, and air takes its place. Gravitational forces remove water only until an equilibrium is reached at which capillary forces (the force of cohesion between water molecules and the force of adhesion between water molecules and soil particles) are sufficient to balance gravitational forces and prevent further drainage. As soil water content continues to decrease, mostly as a result of evapotranspiration, soil capillary forces become strong enough to counteract roots' ability to absorb water. Finally the point is reached at which the turf-grass plant wilts because its roots are unable to absorb the remaining water, which is held too tightly by the small pores. If the soil is not then recharged with rain or irrigation water, the turf-grass plant may die.

One of the main objectives in turf-grass irrigation is to balance soil capillary forces (soil suction) and root water-absorption forces (root suction). The amount of irrigation should be such that the soil always maintains enough suction to prevent water drainage but not enough to counteract root suction and subject plants to water stress.

The ease of achieving this objective depends on soil texture, structure and organic-matter content. Sandy soils are the most challenging. They generally hold only a few per cent of water, which may be used up by turf during a few hours of intensive evapotranspiration, so sands may require irrigation several times a day. Heavy clays are also difficult. They may hold the largest amount of water, up to 30–40%, but very strong capillary forces often prevent most of it from being taken up by roots, so availability of water to plants may be limited. Loamy soils are usually the best. They hold large amounts of water readily available to plants.

Soils with well-developed granular structure hold more water available for plants than unstructural soils. Soil organic matter helps to retain moisture as well.

As mentioned above, after heavy rainfall water moves downwards, passing mainly through the large soil pores. If the soil is porous and uniform, this type of water movement, called saturated water flow, will continue until the large pores are empty. Once that point is reached, movement of water in the soil is described as unsaturated water flow. In such situations, water movement is subject to capillary forces, and flow proceeds in all directions: downward, upward and sideways. These unique properties of soil capillaries can sometimes be used to our advantage. For example, if a layer of sand is placed on top of a layer of gravel, capillary forces in the coarser soil (gravel) are too weak to draw water from the finer soil (sand), and movement will either stop or slow down considerably. This hydrological phenomenon can be utilized in designing soil media for specific turf purposes. For example, in construction of golf course greens and some athletic fields, a layer of sand mixed with a little loam is spread over a layer of very coarse sand or gravel. If a small amount of water is applied, virtually all of it remains 'suspended' in the sandy layer by capillary forces and remains available for growing turf-grass plants. If a large amount of water is applied, for example by heavy rain, its weight will overcome capillary forces in the upper layer and force water into the lower, coarse level, where the large pore spaces allow for quick, unrestricted water drainage. Downward flow will continue until capillary forces once again balance gravitational ones, and then infiltration will stop. In such cleverly modified soils, the very low water-holding capacity of the upper sandy layer can be increased substantially, but very rigorous specifications must be followed. Small differences in texture or thickness of the top layer may substantially change its hydraulic properties. For example, a surface layer that is too shallow may not drain properly; one that is too deep may drain well but result in an excessively dry surface layer.

Except in these very specific situations, where soil layering can be used to our advantage, strongly contrasting soil layers should be avoided. They substantially change hydraulic properties of soil and may consequently negatively influence not only water movement but also soil aeration, microbial activity and numerous other chemical and ecological processes in the soil.

## The Soil Air

As mentioned in Chapter 1, turf-grass roots and most soil organisms require oxygen for respiration. During the respiration process, roots and microorganisms release mostly carbon dioxide, but also some other gases, including toxic ones that may have negative effects on turf and other soil organisms. Carbon dioxide and toxic gases will build up in the soil unless exchange takes place between the air in the soil and the aboveground atmosphere. This process of gas exchange, called *soil aeration,* occurs primarily by diffusion through the soil pores (Fig. 3.10).

The total space not occupied by soil particles in a bulk volume of soil is called the *pore space* and can range from 30% to 70% of the total soil volume. The size distribution of pores in the soil also varies greatly depending on the soil texture and structure. Coarse-textured soils (sands) contain a higher percentage of large pores than do fine textured soils (clays), which in turn contain large numbers of small pores. Large pores help aeration, but small pores are needed for water retention. The ratio of large pores to small pores is therefore considered far more important than total pore space. Numerous studies indicate that the 1:1 ratio that typically exists in sandy loams is the most favourable. Some species tolerate deviation from this ideal ratio better than others, so selection of the most suitable turf-grass species for a given soil is important.

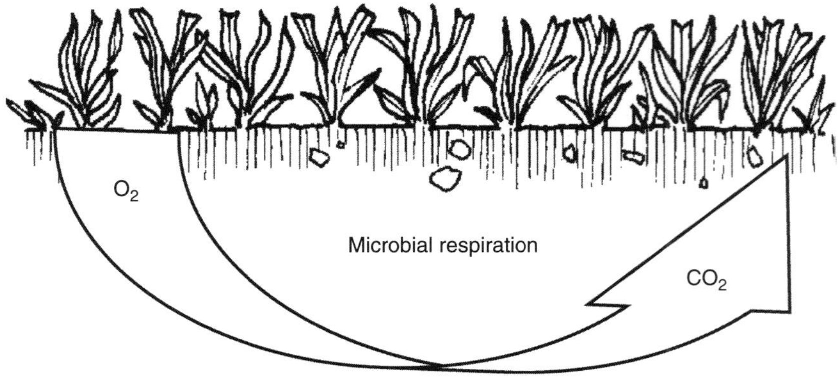


**Fig. 3.10.** Oxygen, carbon dioxide and toxic gases must be exchanged between the air in the soil and the above-ground atmosphere. (Drawing by R. Castro.)

## The Soil Organisms

A fertile soil may contain many organisms, such as microscopic bacteria, fungi, actinomycetes, algae, protozoa and nematodes, as well as relatively large earthworms. The soil organisms perform various activities, mostly beneficial but sometimes detrimental to turf-grass growth and development. They decompose organic matter, fix nitrogen, transform essential elements from one form to another, help in soil aggregation, and sometimes improve soil aeration and drainage. Bacteria are the smallest and most abundant in the soil. Together with fungi and actinomycetes, they contribute substantially to organic-matter decomposition and are mostly beneficial. Particular fungi, nematodes or insects are sometimes pathogenic and unwanted. Some of them infect or feed on roots, shoots or leaves, causing harm to turf-grass plants.

# 4 Turf-grass Establishment

After soil cultivation, turf-grass should be established as quickly as possible. If the soil is left bare, it will be subject to water and wind erosion and general decline of its best properties. In addition, slow or delayed turf-grass establishment provides opportunities for weed encroachment.

Soil should be cultivated and in some instances modified before planting. Proper preparation of the seedbed, for which turf usually corresponds to the top 10–15 cm of the soil surface, is required for the quick establishment of uniform turf. The goal is to create a fertile homogeneous root zone with acceptable infiltration, aeration and drainage.

Depending upon the scope of the project, advance planning may be of relatively little or of very great importance. Establishment of a small residential lawn may require only general gardening skills and some common sense. At the other extreme, establishment of a golf course will probably require input from many people. Golf course construction will almost certainly involve substantial soil modifications, subsurface drainage, installation of an underground irrigation system, construction of artificial ponds and other preparation. For such large projects, detailed sets of drawings, lists of specifications, construction schedules and an overall coordination plan must be developed.

In general, turfs can be established from *seeds* or *vegetatively* from other living parts of the plant. Vegetative establishment refers to any of four basic methods: *sodding*, *plugging*, *stolonizing* and *sprigging*. Choice of the most appropriate of these methods depends on the turf-grass species and the particular situation. Regardless of the method chosen, the steps in site preparation are almost identical.

Steps in producing the optimum root zone for turf establishment may include any or all of: (i) control of existing weeds; (ii) removal of rocks and rubble; (iii) rough grading; (iv) surface and subsurface drainage; (v) soil modification; (vi) fertilization and liming; and (vii) final soil preparation.

## Control of Existing, Persistent Weeds

Hard-to-control weeds should be the primary target at this stage. Many weeds with extensive root systems such as nutsedge (*Cyperus* spp.) are hard to control after grass has been established, so they should be eliminated before establishment. At this stage chemical control with herbicides or soil fumigation are necessary. Non-selective herbicides (see herbicides in Chapter 11, this volume) such as glyphosate could be used to eliminate all growing vegetation. Glyphosate virtually eradicates all plants that come in contact with it. Because it enters the plant only through leaves or stems, glyphosate does not have residual (long-lasting) activity in the soil. Some plants may escape the initial application and require a second application several weeks later. Two other herbicides, amitrole or delapon are selective. They have little effect on broad-leaved weeds but can be used to control undesirable grasses. Unlike glyphosate, amitrole or delapon are taken up through the roots and remain active in the soil for about 6 weeks, so planting should be delayed beyond that period. Because weed control with herbicides is most effective when chemical control is combined with mechanical control, cultivation such as ploughing or disking should follow shortly afterward.

Much less common but much more effective than herbicides is soil fumigation with methyl bromide or other chemicals of similar properties. Fumigation eliminates not only weeds but insects, diseases and nematodes as well. Fumigation can be hazardous, so it should be carried out only by trained and certified professionals. The chemicals used in these procedures are usually toxic and very dangerous if mishandled. Fumigation is not recommended for residential lawns. It is also expensive and is therefore seldom used on large areas.

### Removal of rocks and rubble

Rocks and rubble should be removed from the soil surface after cultivation. Unfortunately, contractors occasionally bury construction debris such as cement, wood and tree stumps. Buried wood eventually decays, but large pieces of metal or cement could pose an ongoing problem (Fig. 4.1). If discovered during cultivation, they should be removed.

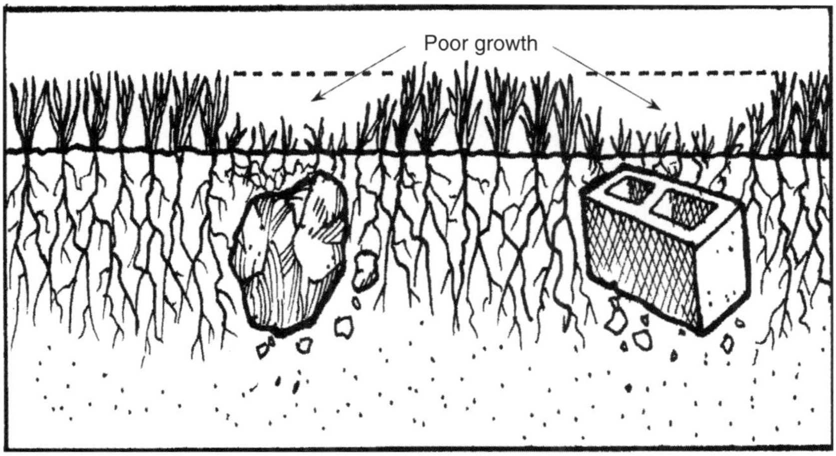


**Fig. 4.1.** Buried debris can restrict turf growth and should be removed before any attempt to establish turf grass.

Large rocks should also be removed from the root zone. As a general rule, a few rocks smaller than a golf ball or the size of a small chicken egg usually do not interfere with turf performance or maintenance and can be left in place. In some situations, however, especially on recreational or sport turfs, the presence of even a few small rocks is not acceptable. They may cause improper operation of, or damage to, specialized turf machinery. In such cases they must be removed before planting.

## Rough Grading

In many cases, the terrain chosen for turf establishment is not even. Especially over large areas, some spots are either lower, higher, flat or sloped. The objective of grading is to provide a relatively smooth, firm surface, which assures both pleasant appearance and adequate drainage of surface water.

For establishment of a residential lawn, rough grading will be needed occasionally. Most experienced contractors build the house at the highest point on the property and provide a 1–2% slope away from the house. Larger turf areas, such as parks or athletic fields, frequently need rough grading. Golf courses almost always need large-scale rough grading. The first step in rough grading is usually removal of the topsoil, which is then stockpiled somewhere nearby for future redistribution over the newly graded site.

An important rule for shaping a subgrade on which topsoil is to be placed is that the subgrade should always follow the finished grade

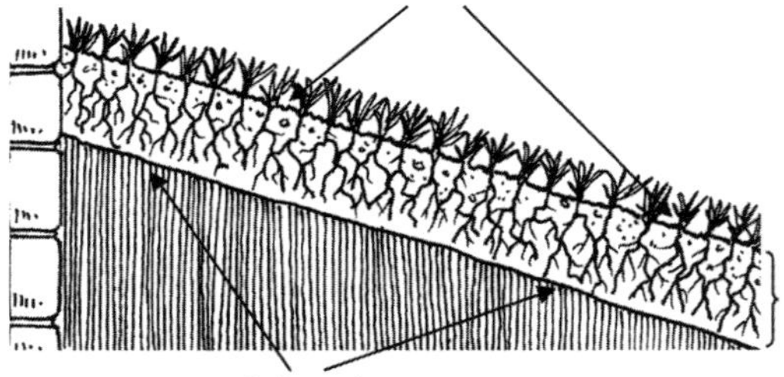


**Fig. 4.2.** The subgrade should conform to the finished grade. (Drawing by R. Castro.)

(Fig. 4.2). If a 2% slope is planned for the finished area, the same 2% should be established in the subgrade. If this rule is not followed, the water that infiltrates through the topsoil may accumulate at the surface of the subsoil, forming puddles below the soil surface. If it accumulates close to the building foundation, it can cause serious damage. Another important rule is that the subgrade should be lowered to allow for topsoil placement. Because topsoil should be at least 10 cm thick, the subgrade should be at least 10 cm below the intended final grade. Steep slopes should also be avoided because of their susceptibility to water erosion and the difficulties of establishing and mowing turf. If the original contour includes a rapid change in elevation, the use of a retaining wall may be the only solution (Fig. 4.3). If the slope is moderately steep, a hedgerow of densely planted shrubs or of erosion-control grasses such as vetiver grass (*Vetiveria* spp.) should be considered.

## Surface and Subsurface Drainage

Effective surface and sometimes subsurface drainage can eliminate many potential problems in turf-grass culture. The contours of a turf-grass area should be both aesthetic and functional; they should rapidly remove excess surface water. Home lawns should slope by at least 1% away from the buildings (i.e. should descend by at least 1 m per 100 m of lateral distance). Athletic turf such as football fields should slope about 2% from the centre. Depressions – areas lower than all surrounding turf – should be avoided. Sometimes, despite proper slope, surface drainage is not sufficient. On soils containing large amounts of clay, rainwater may not be able to drain quickly enough by infiltration, and water will pool on the surface. If the soil is poorly drained and impermeable, and the water table is frequently less than 1.5 m from the soil surface, the area is likely

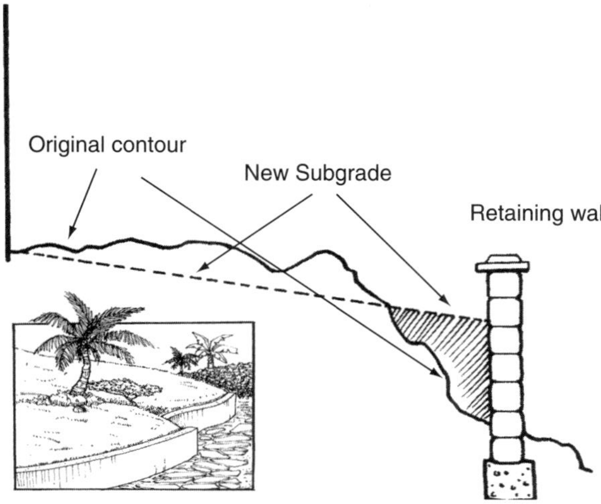


**Fig. 4.3.** If the original contour includes a rapid change in elevation, the use of a retaining wall may be the only workable solution. (Drawing by R. Castro.)

to flood periodically, so a drainage system will probably be needed. Properly designed drainage systems should ensure rapid removal of water from the turf-grass root zone. Factors such as soil texture, topography of the area and desired removal rate should be taken into consideration. Small areas are often drained by perforated plastic tubes buried below the surface. For large-scale projects, professional drainage designers should be consulted. Perforated plastic tubing has become very popular for subsurface drainage of turf-grass areas, but conventional clay or cement drains should also be considered (Fig. 4.4). Clay drains last for decades but are much more expensive to purchase and to install.

**Fig. 4.4.** Perforated plastic and clay drain pipes.

Drainage systems usually consist of small subsurface channels, called *laterals*, arranged in parallel lines connected to larger main channels called *collectors*. Laterals are usually 8–10 cm in diameter. The proper diameter of the collector depends upon the number of laterals connected to it and the size of the area from which water must be removed. To provide unrestricted water flow, lines should slope by at least 1%. Collector lines should be installed at depths beginning at 100 cm below the soil surface and generally should not go deeper than 150 cm. Smaller laterals should start at the depth of 60–70 cm, and the depth should gradually increase towards the collector line. Because a lateral's depth must never exceed that at which it meets the collector, the length of the laterals should not exceed 40 m at the top of the collector and 90 m at the bottom. The distance between lateral drains usually ranges from 5 to 15 m, depending on soil permeability and the other conditions mentioned above. Lines are typically designed in a herring-bone pattern or sometimes in a parallel arrangement (Fig. 4.5). Three critical rules govern design and installation of drainage systems. The primary rule is that water flows only downhill. Second, the terrain and other topographic conditions must allow for adequate outlet of the collected water. Finally, the subsurface drainage must be installed before installation of the irrigation system (if an underground irrigation system is planned). The second rule is the most often forgotten, and the resulting error is the most expensive to correct.

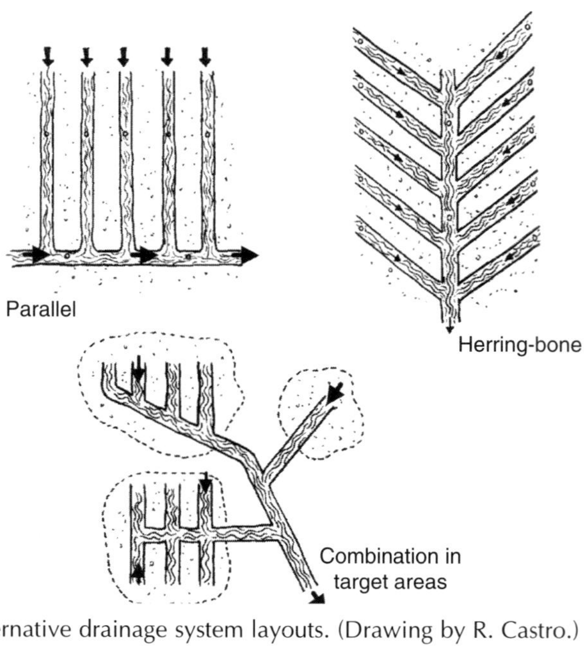


**Fig. 4.5.** Alternative drainage system layouts. (Drawing by R. Castro.)

## Soil Modification

If topsoil was removed before rough grading, it should be brought back and redistributed over the surface. If the topsoil is of poor quality and contains: (i) a high percentage of clay and a low percentage of sand and loam, (ii) a very high percentage of sand and very little loam or clay or (iii) very little organic matter, then high-quality topsoil from another site should be brought in and spread over the surface in a layer at least 10–15-cm thick. A quite popular practice among homeowners – spreading 1–2 cm of highly organic soil over the poor soil or mixing in a similar amount – simply wastes time and money and is not recommended. Such a small amount is insufficient to change soil conditions. At least a 10-cm-thick soil is required to accomplish favourable changes in water retention, aeration, cation exchange capacity and infiltration. After the topsoil is redistributed or modified, it should be allowed to settle undisturbed for about 1–2 weeks. Therefore, if an underground irrigation system is to be installed, the installation should take place before redistribution of topsoil.

### Application of fertilizer and lime

A soil test should be done to determine pH and lime requirements. If lime is needed, it should be applied at this stage and mixed with the soil before seeding or planting. Lime does not move with water, so later applications on established turf-grass are not very effective. The same rule applies to phosphorus. If it is not mixed with the soil before planting, later applications may not be effective. Only a soil test can determine precisely how much of each nutrient should be applied, but application of phosphorus and potassium at 1–2 kg per 100 $m^2$ (100–200 kg/ha) is generally sufficient. Timing of potassium application is less critical, but nitrogen should be applied only 1 or 2 days before planting. Nitrogen applied earlier can be lost to leaching and volatilization (see Chapter 5, this volume) and can serve to promote the growth of competitive weeds.

## Final Soil Preparation

Ideally, final soil preparation should take place just 1 day before planting. The goal is to prepare the best possible environment for the new turf-grass plants. The final seedbed should be firm, moist but not wet, and free of clods, stones and other rubble. If possible, it should be lightly tilled less than 24 h before planting or seeding. If weeds have germinated since the last tillage, a light application of glyphosate can be used. Application of 25% of the label-recommended amount is usually sufficient to control young seedlings. Glyphosate can be applied as little as several hours before planting. If a strong monsoon is expected, final soil preparation should be postponed until the weather stabilizes.

## Seeding

Most of the warm-season turf-grasses are propagated vegetatively, but centipede grass, common cynodon, some types of zoysia grass and several other less commonly used species can be seeded as well. In cool climates, seeding is an essential method of turf establishment, but in warm climates seeding is optional. As a consequence, seed selection is much more limited. Selection of high-quality seed that is adapted to the site conditions and the intended use of the turf is very important. The purchaser of turf-grass seed must study the label to determine the kind, amount and quality of seed in the container as well as to determine how much seed will be needed to establish a particular area of turf. Species differ substantially in seeding rates. For example, common cynodon or zoysia grass may require 1000 g of seed per 100 $m^2$, whereas centipede grass may require only 200 g.

In the tropics the most important characteristic of seeds is their germination rate. Desired germination rates of selected turf species are presented in Table 4.1. The numbers serve as a guide and may differ for different varieties, producers and countries. Accurate rates for particular varieties can be found on the seed label, the purchaser must remember that the germination tests performed by the seed producers took place under controlled conditions, usually more favourable than actual field conditions. In addition, the rates on the seed label take no account of conditions during shipping and storage, which can damage the seed. Seed purity is less important in the tropics than in cooler climates because

**Table 4.1.** Seeding rate and desired minimum germination rate of selected warm-season turf-grasses.

| Species | Seeding rate (kg/100 $m^2$) | Desired minimum germination rate (%) |
|---|---|---|
| Centipede grass | 0.1–0.25 | 70 |
| Cynodon | 0.5–1 | 85 |
| Zoysia | 0.5–1.5 | 50 |
| Carpet grass | 2–3 | 85 |
| Bahia grass | 2–5 | 70 |

of differences related to weed pressures, which will be discussed further in Chapter 11.

Storage temperature is very important in the tropics. In many equatorial countries, metal ocean-freight containers are used for storage, but temperatures in such containers exposed to the sun can easily reach 60°C, which destroys turf-grass seed. Seed should never be stored in metal containers or anywhere else where such high temperatures may occur.

Turf-grass seed should be purchased only from reliable sources and should retain the seed label supplied by the producer (Fig. 4.6). The label may vary slightly in design, but it always supplies essential information about species, variety, germination rate, count of weed seeds, impurities, seeds origin, year of production and the producer's name.

## Seeding time

In tropical climates seeding can take place at any time of the year. In the humid tropics, seeding during the peak of heavy rains should be avoided because seed can be lost to run-off. In the dry tropics, the periods of driest weather should be avoided unless dependable irrigation is available.

## Seeding rate

Seeding rate depends upon turf-grass species, germination rate, germination conditions and the desired rate of establishment. Table 4.1 presents

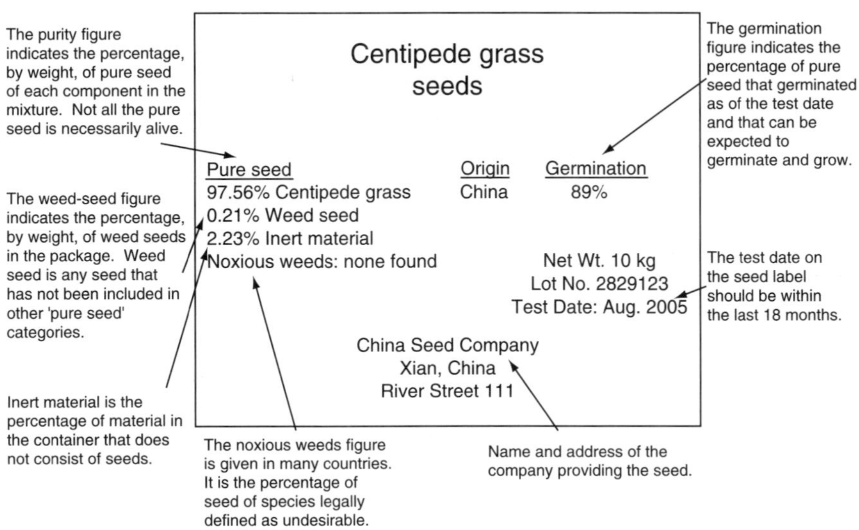


**Fig. 4.6.** Interpreting seed labels.

approximate seeding rates for selected turf-grasses (rates are usually printed on the seed label as well). In the tropics, especially the humid tropics, seeding rates in the upper range are usually optimal. High seedling density during the establishment period reduces weed pressures and produces more competitive turf-grass plants. On the other hard, seeding rates that are too high can extend the period of seedling immaturity, and delicate, juvenile seedlings are less able to compete with weeds and more easily infected by diseases.

### Seeding depth

Most turf-grass seeds are very small and should be planted quite shallowly, at about 0.5-cm depth. Applying the seeds to the surface and lightly rake afterwards can produce the desired planting depth. Seeds planted too deep may not germinate, or their germination may be delayed.

## Seeding Methods

The primary goal is to distribute seeds uniformly. Seeding by hand is popular on small areas. Push-type spreaders, both broadcast and drop types, are used on areas up to 1–2 ha. Tractor-mounted spreaders can be used on larger areas. Spreaders used for seeding are the same as those used for fertilizing (see Chapter 5, this volume). They must be calibrated before use (see Appendix).

Overall, seeding warm-season grasses requires more effort than seeding cool-season grasses. The seeds are very small, often slippery and mostly dark in colour, so they are not visible on the ground. Filling the spreader with seeds and applying them directly can be very risky. Small imperfections in the spreader's gate opening may permit considerable loss of seeds before their escape is noticed. The seeds should therefore be thoroughly mixed with dry, preferably white, sand. The percentage of sand is not important, but a proportion of 1 part seeds to 10 parts sand usually works well. The spreader should be calibrated for the seed–sand mixture. The desired amount of seed should be divided in half and applied, with sand, in two passes over the area at right angles to each other. This practice assures uniformity of coverage and prevents accidental skips. The visibility of white sand on the ground reveals any areas of non-uniformity. Small areas such as home lawns can be seeded from a jar with holes punched in the lid (Fig. 4.7). The desired amount of seed can be mixed with five parts white sand. All at once or in several portions, the seed–sand mixture is placed in the jar, which acting like a large salt-

**Fig. 4.7.** A hand spreader for seeding small areas.

shaker, allows for a uniform application of seeds. White sand guides the applicator very efficiently by revealing which spots received more seeds and which received less.

Areas that are difficult to reach, such as steep hillsides or roadsides, are sometimes hydroseeded. Hydroseeding is a specialized type of grass establishment in which seeds, paper pulp and fertilizer mixed with water are sprayed onto the soil surface from a distance by special equipment (Fig. 4.8). The paper pulp sticks to the soil, prevents seed run-off, retains moisture for days without rain, and lasts until the grass is well established. Almost any type of area can be hydroseeded, but the cost is substantial.

## Pre-germination

Pre-germination is used where a very rapid establishment is desired or to reduce weed pressure. Some seeds such as those of zoysia grass and centipede grass germinate slowly, and the resulting lengthy interval between seeding and germination increases the opportunity for invasion by weeds. In the pre-germination process, seeds are soaked in water, usually for 5–10 days. When the first seeds show signs of germination, the water-soaked seeds are mixed with dry sand and spread over the desired area.

**Fig. 4.8.** In hydroseeding, seeds, paper pulp and fertilizer mixed with water are sprayed onto the soil surface from a distance.

## Germination

Seeds need moisture to germinate, but the most critical time comes just after germination, when seedlings have begun to root but have not yet developed a root system. If adequate moisture is not available even for several hours within this period, the entire stand may be lost. The surface must be kept moist, in the absence of rain, by light applications of water several times a day. After this critical period, which lasts for about 1 week, the grass-root system develops to the extent that water can be obtained from the underlying soil. From this time, irrigation should gradually increase in volume and decrease in frequency.

## Mulching

When irrigation during the critical period after germination is not available, mulching (covering with a layer of dried or partially decomposed plant matter) may improve the chances of success. Straw mulching is inexpensive and provides excellent results. The rule is to spread enough straw to cover the soil but to leave the surface somewhat visible. Straw or another similar mulch helps to retain surface moisture. Daily rain or irrigation is usually sufficient. The straw can be raked away several weeks later or left to decompose naturally.

## Post-germination Care

After germination, proper irrigation is the key factor in quick establishment. If straw mulching was used, removal of at least a portion of it is beneficial. Because of wind or water flow, straw tends to accumulate in certain areas, blocking light that is essential for the quick growth of seedlings. Seedlings should be also protected from the effects of traffic. Placing ropes around newly seeded turf is usually effective.

### Mowing

The first mowing should occur when seedlings reach a height one-third greater than the anticipated mowing height. For example, if centipedegrass turf is to be maintained at a 3-cm cutting height, seedlings should be mowed when they reach a height of 4 cm. From this point mowing should continue at standard frequency and should be guided by the same 'one-third rule' (see Chapter 6, this volume). Mowing less frequently removes too much leaf area at one time and can set the plant back severely.

### Fertility

Shortly after the first mowing, a light application of nitrogen may substantially speed up turf establishment. Not more than 0.25 kg N/100 $m^2$ (25 kg/ha) should be applied and watered into the soil.

### Irrigation

As previously mentioned, irrigation should be sufficient to prevent moisture stress in young seedlings. The soil surface should be moist as long as the root system remains poorly developed. Common cynodon develops a fair root system in 3–4 weeks, centipede grass in 4–5 weeks and zoysia grass in 6–8 weeks.

### Weed control

In tropical climates, weeds may invade newly established turf massively. To someone inexperienced, a great number of weeds and few visible turf-grass seedlings may cause fear that turf establishment is failing. In addition, at this stage herbicides should not be used because the young grass seedlings are too sensitive to tolerate them and may sustain severe injury. Fortunately, appearance of the newly seeded turf greatly improves after the first mowing and keeps improving with time. Mowing at the correct height removes the growing points of most weeds without harming the crowns of the turf-grass plants, and in many cases competition from strong turf-grasses weakens weeds, so herbicides may not be needed at all. Cynodon is an exception; in the humid tropics, it is usually not sufficiently competitive and requires herbicides for control of both grassy and broad-leaf weeds. Herbicides such as 2,4D, MCPA, dicamba and MSMA can be quite safely applied after 4–5 weeks (see Chapter 11, this volume).

## Vegetative Establishment

### Sodding

Sod is established turf that is harvested, with roots and soil attached to it, and transplanted from its place of origin and installed like carpeting to grow in another place. Sod is usually produced on a farm that specializes in commercial production of turf-grass. Sodding is the most expensive method of turf establishment but produces an established turf within hours rather than weeks or months. Almost all warm-season turf-grasses can be sodded, but zoysia grass, St Augustine grass and centipede grass are established in this way most often. The only disadvantage of sod is its high cost. Otherwise it provides an instant turf, which requires no extra maintenance other than daily watering during the early weeks of establishment.

After being harvested at the farm, sod is brought to the site in the form of rectangular pieces or as long rolled strips. Properly harvested sod is delivered to the site with a thin layer of soil. The general rule is to harvest a layer of sod as thin as it can be cut and still hold together. For most species, that means about 2-cm thick. After being cut, the sod should be moved immediately to its destination and laid as soon as possible. Sod that has been stacked and left for several days usually dries out and often turns yellow or a pale green colour. In tropical climates, stored sod loses its vigour after 2–3 days. After 5–6 days stockpiled sod is usually of very little value, even if watered.

Sod should be laid in a bricklike pattern, such that the corners of four pieces never come together at one place (Fig. 4.9). Gaps between pieces should be as small as possible to prevent invasion by weeds. On hills sod should always be laid perpendicular to the slope. Where the slope is

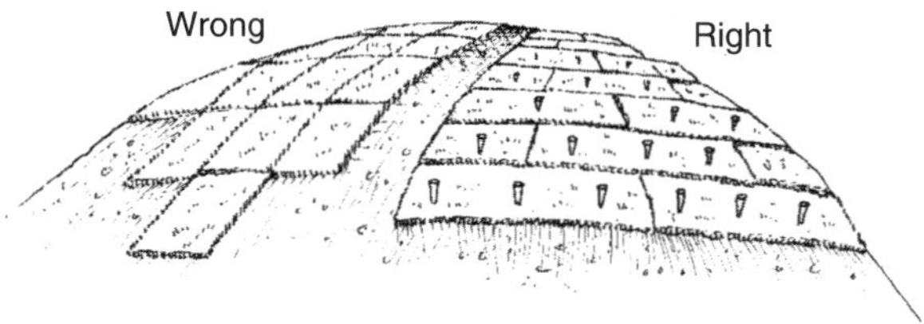


**Fig. 4.9.** Sod should be laid in a bricklike pattern. On hills, rows should be laid perpendicular to the slope and pinned to the ground. (Drawing by R. Castro.)

steeper than 10%, sod movement should be prevented by wooden stakes used to pin it to the ground during the 2–3-week rooting period. Sod should be well watered immediately after being installed. Lighter watering should continue until rooting is obvious, usually for about 2 weeks. Sometimes sod farms use highly organic soil as a rooting medium. This type of soil absorbs water like a sponge, and if only slightly wetted, it holds water on the surface without allowing its infiltration into the underlying soil. For that reason incompatible soils should be avoided if possible. Otherwise, the soil under new sod should be periodically checked (the corners of selected sod pieces can be curled back for the purpose) and kept continuously moist but not excessively wet. About 2–3 weeks after sodding, 0.25–0.5 kg/100 $m^2$ of nitrogen fertilizer should be applied. The first mowing should take place as soon as the turf can tolerate traffic without splitting apart. In most cases, within 4–5 weeks the sodded area can receive maintenance similar to that suitable for mature turf.

## Stolonizing

Every bud on a stolon, just like every seed, can potentially become a new plant. Stolonizing is spreading pieces of stolons rather than spreading seeds. Many tropical (warm-season) turf-grasses do not produce viable seeds or, like most hybrids, do not produce seeds at all. For these grasses, stolonizing is the primary method of turf establishment. Almost all tropical grasses can be stolonized, but hybrids of cynodon and several species of zoysia grass are stolonized most often. Stolons are commonly harvested from mature turf with a vertical mower (see Chapter 6, this volume). Core cultivation (see Chapter 8, this volume) is another excellent source of high-quality stolons. Besides stolons, cores include roots and soil that can keep them alive much longer. Unfortunately, cores are quite heavy and bulky, so they can be used only in a nearby location. Stolons are light, easy to pack, and if refrigerated stay viable for 2–3 weeks. Stolons can be transported to distant locations, even to other continents.

After they are harvested, stolons should be kept moist. In most situations, stolons are broadcast on the day of harvest or the following day. On large areas special machines are used, but on small areas the distribution can be done manually. Depending on turf-grass species, condition of the harvested stolons, desired rate of establishment and the level of post-planting care available, the amount of plant material needed to cover 100 $m^2$ areas ranges from 75 to 150 l of loosely packed stolons. Another way to estimate the quantity of stolons needed is to use the sod-stolon multiplication factor. For cynodon and zoysia grass this factor is around 10; it means that 1 $m^2$ of sod must be harvested to stolonize a 10 $m^2$ area.

After broadcasting, stolons should be covered by soil. Top-dressing (see Chapter 8, this volume) with 3–5 mm of soil or sand is usually sufficient to prevent excessive desiccation and possible movement of small plant pieces by the wind. Stolons must not be buried completely – approximately 15–25% of each stolon should extend above the soil surface. Light rolling is also recommended, because it improves the plant-to-soil contact. The stolonized area should be watered and maintained in a way similar to that for a seeded area (see sections on Germination and Post-germination Care, above).

## Sprigging

Sprigging is a modification of stolonizing. It involves placing stolons in narrow furrows spaced 15–20 cm apart (Fig. 4.10). It is performed mainly

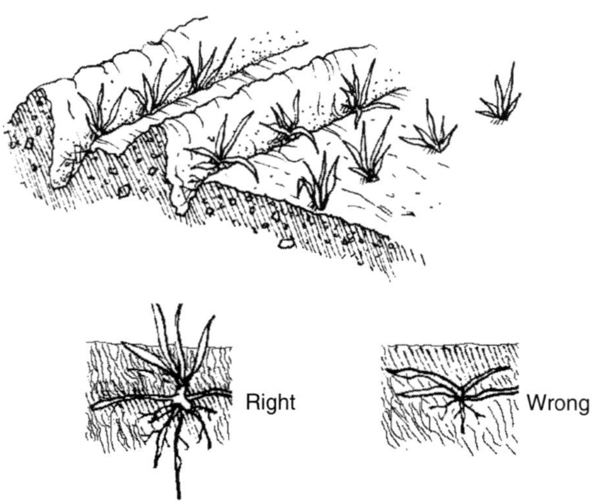


**Fig. 4.10.** The ideal placement of sprigs leaves one-quarter of each sprig above ground after planting in the furrow. (Drawing by R. Castro.)

on smaller areas and often with species having larger stolons, such as St Augustine grass. A stolon should be placed every 5–10 cm in the furrow and manually covered by soil such that 15–25% extends above the surface. Major advantages of this method are minimal loss due to desiccation, the need for only half as much planting material or less, and less rigorous post-planting care. The major disadvantage is higher labour costs.

## Plugging

Only stoloniferous or rhizomatous turf-grasses can be established by plugging. Because almost all warm-season turf-grasses belong to this category, plugging can be widely used in tropical climates. Plugging, sometimes called spot sodding, is the planting of small pieces of sod spaced apart over a large area (Fig. 4.11). The small pieces of sod, called plugs, are commonly formed by cutting of sod strips into small square pieces. Establishment by plugs is chosen most often for small areas and involves a considerable amount of manual labour. Plugs are usually about 5 × 5 cm and are placed in a grid pattern on 15–30-cm centres. Small holes are punched in the soil with a stick made of wood or metal,

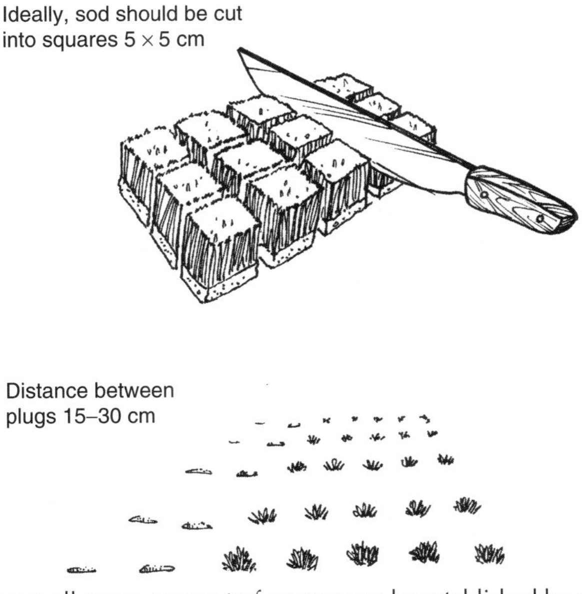


**Fig. 4.11.** Almost all warm-season turf-grasses can be established by plugging. Ideally, sod should be cut into squares 5 × 5 cm, which are then planted 15–30 cm apart. (Drawing by R. Castro.)

and each plug is placed in a hole and pressed down, usually with the planter's foot. Major advantages of plugging are low costs of planting material and minimal losses. Major disadvantages are high labour intensity and the possibility of weed invasion during the period before the plugs spread to fill the empty space. Another disadvantage is slow establishment. Fast-growing cynodon may completely cover the soil within 4–5 months, but slow-growing zoysia grass may take more than a year. Unlike seeding or stolonizing, plugging in the tropics is often accomplished over an extended period of time. Sometimes, because of the lack of commercial sod or more often because of its high cost, many residential lawns are gradually established by the continuous transplanting of sod pieces from recently established areas to the new areas. Plugging's multiplication factor, which ideally ranges between 15 and 20 is an important advantage of this practice. In most cases only 1 $m^2$ of sod is needed to produce 5 × 5-cm plugs for a 15–20-$m^2$ area. If the size of the plugs is reduced to 2 × 2 cm, as it often is, 1 $m^2$ of sod can cover as much as 100 $m^2$ of soil.

*Strip sodding* is a modification of plugging. Instead of plugs, strips of sod are laid in long rows 15–20 cm apart. As in plugging, the bare areas are filled in by spreading stolons and rhizomes. This method of turf-grass establishment is somewhat faster, is often performed on larger areas, uses more plant material, but involves much less manual labour. Zoysia grass is the species most often established by strip sodding.

## Turf Renovation

Sometimes mismanaged turf is in such poor condition that reasonable improvement is not achievable through routine cultural practices. Sometimes turf-grass species must be changed to match changing conditions, e.g. on residential lawn progressively shaded by growing trees. In cases like these, if slopes and grades on the site are adequate and no major soil modifications are needed, renovation might be considered. Renovation is the replacement of an existing turf with a new one without most otherwise essential establishment procedures.

The first step is killing the existing turf and weeds on the area subject to renovation. This process usually requires a nonselective herbicide such as glyphosate. Glyphosate is taken up by leaves within several hours and translocated into roots, rhizomes, stolons, tubers, etc., which it kills slowly over the next 7–10 days. Some tough weeds, such as nutsedge, may require an additional application of glyphosate 1 week later. After the turf area has been sprayed with herbicide, depending upon the establishment method chosen (seeding or vegetative propagation), several alternative steps can be carried out.

In tropical climates, renovation by seeding is performed only occasionally. The process is quite difficult and usually costs more than vegetative renovation. Nevertheless, if seeding is chosen, the area being renovated should be mowed as low as possible, preferably to a height of less than 1.0 cm. Reducing the height of the previous turf ensures adequate light for the emerging seedlings (the old turf should not be mowed before application of glyphosate, because this herbicide is taken up by leaves, and mowing would reduce its effectiveness). The next step depends on the amount of thatch that has accumulated in the area (see Chapter 8, this volume). The seeds must germinate in contact with the underlying soil. Seeds that germinate in the thatch are unlikely to survive. Thatch layers more than 3-cm thick should be removed with a sod cutter and discarded. For thinner thatch layers, a vertical mower (dethatcher) should be used to slice through the sod to the underlying soil in two directions, and the excess thatch should then be raked from the surface. Core aerification may also be advisable if the soil is compacted. Once the thatch has been removed, the seed can be spread on the area being renovated. Because some seed will germinate in the debris above the soil surface and be lost, the seeding rate appropriate for bare soil should be increased by at least 25%. The seed should be worked into the seedbed as much as possible; an additional pass with the vertical mower will bring the seed into better contact with the underlying soil. A starter fertilizer should be applied to the surface as a final step at the same rate recommended for normal seeding. The area should be irrigated as would any other newly seeded area. The debris from the old turf cover will serve as mulch and may reduce the need for irrigation. Post-germination care should be the same as that for any other seeded area.

If the new turf in the renovated area will be established vegetatively, several other approaches can be used. If sodding is chosen, a thin layer of soil should be distributed over the area, and sod should be laid down in the same manner as for sodding on bare soil. The purpose of the soil layer is to provide continuity between the sod and the underlying soil. Without this step, sod placed on a thick mat of dead grass will suffer from desiccation. The added soil also provides a better environment for microbial activity and therefore speeds up decay of the old turf.

If plugging is chosen, the turf area should be mowed the day after application of glyphosate. Mowing should be at the usual height, and the clippings should be collected and removed. Mowing should be as accurate and visually appealing as possible because, even though the old grass is dead, it will serve as temporary turf for several months while the plugs spread. Removal of clippings is important because glyphosate speeds up maturation of weed seeds; if it is left on the surface, weeds may germinate and contaminate future turf area. When the grass turns yellow (usually within a week) but some weeds still remain green, a second application of glyphosate may be needed. The next step is to punch holes and insert plugs. In contrast to plugging into bare, cultivated soil, the holes must be punched with some heavy device, such as a metal crowbar, able to penetrate at least 4–5 cm into the soil. After the plugs are inserted, the area should be cared for as would any other plugged area.

The author once observed an unconventional turf-renovation method at a tropical Asian resort, where cynodon turf area was converted to St Augustine grass. A core aerification machine (see Chapter 8, this volume) punched holes in already brown, dead turf, removing soil cores to the surface. Workers manually inserted pieces of St Augustine stolon into each individual hole and then filled the holes with soil taken from the surface. The results were exceptional. Within two months, the cynodon turf area was entirely converted to St Augustine grass. This type of renovation requires a great deal of manual labour and is achievable only where labour is inexpensive. Otherwise, sprigging and stolonizing cannot be used to renovate turfs. Nearly all planting material will be lost to desiccation.

Renovation, unlike regular turf establishment, can be performed at the peak of the rainy (monsoon) season. In most cases, heavy rains cause no harm.

# Turf Nutrition and Fertilization

5

## The Essential Nutrients

Unlike animals, which must digest proteins, carbohydrates, fats and other organic nutrients, plants require only inorganic mineral nutrients. Sixteen chemical elements have been identified as necessary for the growth of turf-grasses: carbon (C), hydrogen (H), oxygen (O), nitrogen (N), phosphorus (P), potassium (K), calcium (Ca), magnesium (Mg), sulphur (S), iron (Fe), manganese (Mn), zinc (Zn), boron (B), copper (Cu), molybdenum (Mo) and chlorine (Cl). These are the *essential nutrients*, those that are required for growth and development of turf and without which the grass plant cannot complete its life cycle. Carbon, hydrogen and oxygen are taken from the air and water. These elements play an important role in photosynthesis, the process of forming carbohydrates (i.e. compounds consisting only of carbon, hydrogen and oxygen) from carbon dioxide ($CO_2$) and water ($H_2O$). The other 13 essential elements are categorized into two groups: *macronutrients* and *micronutrients*. Nitrogen, phosphorus and potassium are frequently called primary macronutrients. Turf requires them in the largest quantities, and they are usually the first in which the plant becomes deficient. The secondary macronutrients – calcium, magnesium and sulphur – are just as important, but they are required in smaller quantities and seldom produce symptoms of deficiencies. The micronutrients – iron, manganese, zinc, boron, copper, molybdenum and chlorine – are required by plants in only very small quantities. Iron is the only micronutrient in which turf-grass plants are likely to become deficient.

 

Percentages of elements present in a typical grass plant are presented in Table 5.1.

## Nitrogen

Unlike other essential plant nutrients, nitrogen is not found in significant amounts in the rocks and minerals of the earth's crust, but most of the nitrogen in soil is in organic form and unavailable to plants. Fates of nitrogen in turf are shown in Fig. 5.1.

Soil organic matter contains about 5% nitrogen and must break down to release it to the soil. In the tropics, organic nitrogen is continuously converted to its inorganic forms through the process of *mineralization*. A large part of the mineralized organic nitrogen undergoes a process called *nitrification* and is finally converted to the soluble nitrate form ($NO_3$). Besides the nitrates, which originate from mineralization of organic matter, inorganic nitrates are applied directly during fertilization. Because they exist in soil solution as negatively charged ions, they are not held by negatively charged soil colloids and can easily be removed from the root zone by leaching.

The rate of nitrogen mineralization depends on microbial activity; so soil moisture, temperature, pH and aeration directly affect the process. In the tropical climate, substantial amounts of nitrogen are lost before they can be taken up by roots.

Although they are valuable to the turf, nitrates are considered dangerous water pollutants. In the tropics, if amounts are applied that are too

**Table 5.1.** Percentages, by weight, of elements in a typical turf-grass plant.

| Element | Percentage of dry weight |
|---|---|
| Oxygen | 44.0 |
| Carbon | 44.0 |
| Hydrogen | 6.0 |
| Nitrogen | 1.5 |
| Silicon | 1–2 |
| Potassium | 0.90 |
| Phosphorus | 0.20 |
| Sulphur | 0.15 |
| Calcium | 0.25 |
| Magnesium | 0.20 |
| Chlorine | 0.15 |
| Iron | 0.10 |
| Manganese | 0.05 |
| Other | 0.5–1.5 |

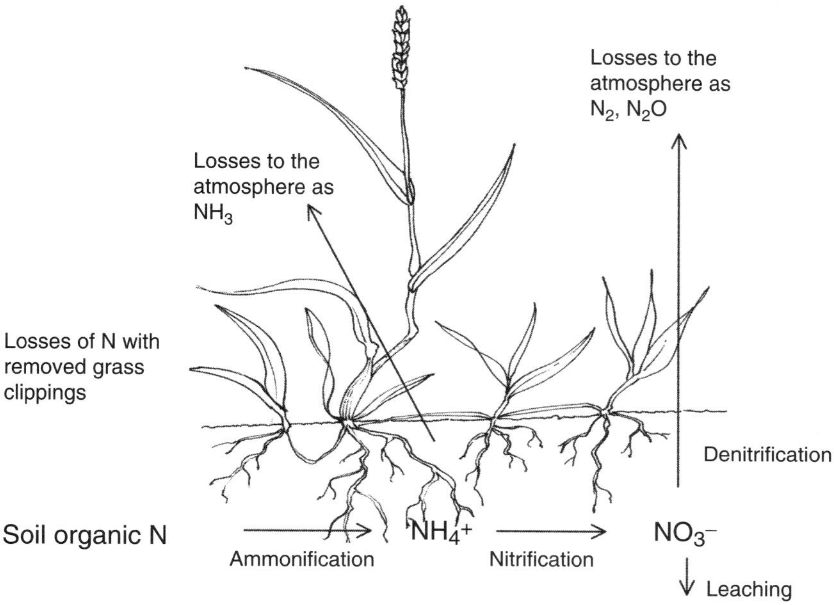


**Fig. 5.1.** Fate of nitrogen in turf.

large for the plants to absorb immediately, heavy monsoon rainfalls can leach them from the soil into the ground water. Techniques designed to minimize this process are use of slow-release fertilizers and 'spoon-feeding', the frequent (weekly) application of small amounts of fertilizer rather than infrequent applications of larger amounts.

In addition to leaching, nitrogen can be lost from the soil in gaseous form. Organic matter and, to an even larger extent, urea and other fertilizers containing ammonia ($NH_4^+$) undergo the process called *ammonification*. Ammonium ions valuable for plants become converted to ammonia gas ($NH_3$) and lost to the atmosphere. In the rainy season, when turf remains flooded for long periods, nitrate fertilizers may be lost through a process called *denitrification*. Under anaerobic conditions, where oxygen is limited, nitrate ions can be biologically reduced to nitrous oxide ($N_2O$) or nitrogen gas ($N_2$) and lost to the atmosphere. For economic and environmental reasons, therefore, nitrogen applications must be managed carefully.

## Phosphorus

Phosphorus forms insoluble compounds with aluminium and iron at low soil pH and with calcium at high soil pH. Even in neutral soil, phosphorus is almost always combined with other elements. Insoluble compounds slowly change into soluble orthophosphates ($HPO_4^-$ and $H_2PO_4^-$), which can then be taken up by plants. Unlike the compounds of nitrogen, amounts of soluble forms of soil phosphorus are small, its mobility in soil is low and very little is lost to leaching or other processes. Unlike many other crops, turf-grass does not need high levels of phosphorus in the soil. Phosphorus deficiency mainly affects the generative parts of the plant, such as the seed-head and seeds, which are not important in turf. In fact, a slight deficiency is beneficial, because low levels of phosphorus hold back production of seed-heads, which are unsightly and undesirable in turf. Because symptoms of phosphorus deficiency seldom appear in leaves, turf managers pay little attention to the level of soil phosphorus.

## Potassium

Potassium is involved in growth processes and strongly influences stress tolerance, especially cold-temperature tolerance. Researchers disagree, but most evidence suggests that potassium is required by turf in quantities about a quarter to a half those of nitrogen. Tropical grasses may require less potassium than cool-season grasses, but not enough research has been done to confirm this theory. Potassium ions are released to the soil very slowly during the process of physical and chemical weathering of minerals. Some soils contain large quantities of potassium, and others very little. A portion of soil potassium becomes trapped or fixed between layers of clay particles and is called the soil reserve. Potassium is usually readily available in the soil solution or in an exchangeable form on the soil colloids. Potassium occurs as a positively charged cation ($K^+$) and usually constitutes a large portion of the soil's cation exchange capacity. Potassium is relatively mobile but not as mobile as nitrogen. Management of potassium is more critical than management of phosphorus but still far less essential than management of nitrogen.

## Calcium

Calcium exists in the soil as positively charged cations ($Ca^{2+}$). Most soils contain enough calcium to support growth of turf-grasses. Calcium is applied (a process known as liming) to increase low soil pH rather than to support plant growth. Calcium deficiencies in plants are extremely rare.

## Magnesium

Magnesium, like calcium, exists in the soil as positively charged cations ($Mg^{2+}$). Deficiencies occur sometimes, especially on coarse-textured sandy soils. Magnesium deficiency is frequently associated with low soil pH.

### Sulphur

Sulphur deficiency is infrequent. Sandy soils are most likely to produce sulphur deficiency, especially if low in organic matter.

Plants are seldom deficient in micronutrients such as copper, boron, manganese, zinc and molybdenum. Iron is the only micronutrient in which plants are frequently deficient, especially in alkaline (high-pH) soils. Soil test procedures in which iron availability is determined are usually reliable, but occasionally, for unknown reasons, such a test may indicate serious iron deficiency in soil on which the turf shows no symptoms. Likewise, the test may sometimes indicate iron to be sufficient while the plants clearly show deficiency symptoms. The best way to confirm iron shortage is to apply it to a small area and to observe how the turf responds.

Micronutrients are seldom applied in turf-grass management. Although they may be required under very unusual circumstances, in general, their use should be avoided. Most micronutrients can cause injury to the leaves and generally do not serve any useful purpose. Applications of micronutrients should always be based on confirmed observations of deficiency symptoms in the field, soil tests and tissue test results. Soil pH is an important tool for managing micronutrients. Exception for molybdenum, levels of plant-available forms of these elements decrease with the rise of pH.

## Fertilization

Anyone who has ever grown a plant understands the importance of fertilization. Application of some fertilizers makes plants grow faster, but applying too much can harm or kill them. Different turf-grass species have different nutrient requirements just as different levels of maintenance require different nutrient management. A single fertility programme cannot be designed to benefit all types of turfs and all management regimes.

Every fertility programme should start with a soil analysis and, if possible, a leaf-tissue analysis. The governments of most countries around the world provide these services for a relatively modest fee. No one should consider saving money by not doing a soil test. The cost is insignificant compared to the losses that can result from the lack of adequate soil information. A leaf-tissue test is also very helpful.

It provides specific information about the types and amounts of nutrients being taken up by specific turf species, of which the soil test provides only a rough idea. Combining soil and plant-tissue tests provides a complete picture of soil fertility. Leaf tissue should always be tested before establishment of a more complex turf area, such as a golf course, athletic field, tennis court or even playground.

The results of a soil test can only be properly interpreted if the soil sample they are based on was representative. Soil samples can be taken at any time of the year, but it is better to avoid times shortly after exceptionally heavy rainfall. Samples should be collected with a soil probe or any other tool that is able to cut to a depth of 15 cm. If the soil is uniform in texture and colour and has not received recent applications of fertilizer, organic amendments or lime, 10–15 samples from random locations are sufficient for an area the size of a large lawn, park lawn or football field. All the samples should be placed in a large container and mixed well. Finally, about 0.5 kg of this composite sample should be sent to the laboratory. Separate composite samples should be assembled from areas that differ in soil texture or colour and in areas that have received recent amendments. In many countries, soil test laboratories present results as numerical values that represent sufficiency levels for a particular turfgrass. In other countries, only general sufficiency levels are presented. Tables 5.2 and 5.3 present sufficiency ranges of nutrients in soil and leaf tissue.

**Table 5.2.** Sufficiency ranges of nutrients in the soil.

| | Amount (kg/ha) represented by each soil-test designation | | | |
|---|---|---|---|---|
| Nutrient | Low | Medium | High | Very high |
| Phosphorus | <15 | 15–30 | 30–120 | >120 |
| Potassium | <100 | 100–150 | 150–300 | >300 |
| Calcium | <500 | | | |
| Magnesium | <40 | | | |
| Sulphur | <15 | 15–50 | >50 | |
| Boron | <0.5 | 0.5–1.5 | >1.5 | |
| Copper | <0.5 | 0.5–5 | >5 | |
| Iron | <15 | 15–120 | >120 | |
| Manganese | <10 | 10–50 | >50 | |
| Zinc | <2 | 2–5 | >5 | |

Source: J.B. Jones, Jr., B. Wolf and H.A. Mills (1991) *Plant Analysis Handbook*. Micro-Macro Publishing, Athens, Georgia.

**Table 5.3.** General tissue sufficiency ranges of nutrients for cynodon. Values for other warm-season turf-grasses may differ slightly for macronutrients and considerably for micronutrients.

| Nutrient | Low | Medium | High |
|---|---|---|---|
| | Percentage of tissue composition | | |
| Nitrogen | 2.50–2.99 | 3.00–5.00 | >5 |
| Potassium | 0.70–0.99 | 1.00–4.00 | >4.00 |
| Calcium | 0.30–0.49 | 0.50–1.00 | >1.00 |
| Phosphorus | 0.12–0.14 | 0.15–0.50 | >0.50 |
| Sulphur | 0.12–0.14 | 0.15–0.50 | >0.50 |
| Magnesium | 0.10–0.12 | 0.13–0.50 | >0.50 |
| | Parts per million of tissue composition | | |
| Iron | 40–49 | 50–350 | >350 |
| Manganese | 16–24 | 25–300 | >300 |
| Boron | 4–5 | 6–30 | >30 |
| Copper | 3–4 | 5–50 | >50 |

Source: J.B. Jones, Jr., B. Wolf and H.A. Mills (1991) *Plant Analysis Handbook.* Micro-Macro Publishing, Athens, Georgia.

## Nutrient Uptake

Turf-grass can obtain nutrients through root absorption from the soil solution or by direct foliar absorption. Once absorbed, the nutrients can move passively up through the transpiration stream or be actively transported within the plant. Passively transported nutrients dissolved in the water travel through the xylem to the leaves. Actively transported nutrients move from one cell to another to reach their final destinations. Passive transport does not require energy from the plant. To carry out active transport, the plant must expend some of the energy produced during the process of respiration. Nutrients absorbed by roots are usually transported passively upward through the xylem; those absorbed by turf foliage are transported actively.

Most nutrients are absorbed by roots from the soil. Only occasionally is foliar fertilization, also called foliar feeding, used to supply nutrients through the leaves. Foliar application rates are usually smaller than those intended for absorption through the roots, and the response of the turf is typically faster. Foliar applications are used when nutrients are immobilized in the soil (as iron is in alkaline soils) or when a quick improvement in appearance is needed. Golf course superintendents and sports-stadium managers spray nitrogen or iron solution over turf foliage shortly before tournaments to make grass look greener. It is not unusual for the turf to respond to such treatment within minutes, but under regular management, foliar fertilizers are impractical and seldom used. Dissolved nutrients must be applied often, and if solutions are too concentrated, considerable leaf injury can result.

## Turf Fertilizers

Most inorganic fertilizers dissolve quickly in the soil water and are immediately available to plants for uptake. Plants absorb the majority of nutrients from the soil solution in the form of simple, inorganic ions (Table 5.4). Larger molecules can also be absorbed, but much more slowly.

Organic fertilizers are more complex, require time to be broken down into forms usable by plants, release their nutrients more slowly and progressively meet the demand of the growing plants. An additional benefit is that organic fertilizers add considerable quantities of organic matter to the soil, therefore improving drainage, aeration, water-holding capacity and ability of the soil to hold nutrients.

Many fertilizers are available in the market. Manufacturers offer a wide variety of nutrients, various nutrient proportions, combinations of organic and inorganic, various types of nutrient release, various formulations, and so on. Basic information about each fertilizer is printed on its

**Table 5.4.** Forms in which essential mineral elements are available to turf-grasses.

| Element | Available forms |
|---|---|
| Nitrogen | $NH_4^+$, $NO_3^+$ |
| Phosphorus | $HPO_4^{2-}$, $H_2PO^-$ |
| Potassium | $K^+$ |
| Sulphur | $SO_4^{2-}$ |
| Calcium | $Ca^{2+}$ |
| Magnesium | $Mg^{2+}$ |
| Iron | $Fe^{2+}$, $Fe^{3+}$ |
| Manganese | $Mn^{2+}$ |
| Boron | $B(OH)_4^-$ |
| Copper | $Cu^{2+}$ |
| Zinc | $Zn^{2+}$ |
| Molybdenum | $MoO_4^-$ |
| Chlorine | $Cl^-$ |

PREMIUM

LAWN FERTILIZER

15-5-10

Guaranteed Analysis

| | |
|---|---|
| Total Nitrogen (N) | 15.0% |
| Ammoniacal Nitrogen | 2.9% |
| Urea Nitrogen | 12.1% |
| Available Phosphate ($P_2O_5$) | 5% |
| Soluble Potash ($K_2O$) | 10% |
| Sulphur (S) Total | 3% |
| Iron (Fe) Total | 2% |

KEEP OUT OF REACH OF CHILDREN

**Fig. 5.2.** A typical fertilizer label.

label (Fig. 5.2). Almost every country regulates by law the types of information included on fertilizer labels, but all labels around the world would contain information about the *fertilizer analysis*. The analysis represents the percentage by weight of nitrogen (N), phosphate ($P_2O_5$) and potash ($K_2O$) in the fertilizer. These three numbers are always printed in the same order and usually appear in a larger font than other information.

It is a common misconception that the three numbers represent directly the percentages of nitrogen, phosphorus and potassium in the fertilizer. In fact, only the first, the nitrogen number, represents the true percentage of its element. Because the other two numbers represent compounds of phosphorus and potassium, rather than the elements themselves, the proportions of the elements are different. $P_2O_5$ is only 44% phosphorus, and $K_2O$ is 83% potassium, e.g. a 15-15-15 fertilizer contains not 15% nitrogen, 15% phosphorus and 15% potassium but (15% $\times$ 1.00 =) 15% nitrogen, (15% $\times$ 0.44 =) 6.6% phosphorus and (15% $\times$ 0.83 =) 12.45% potassium.

In most instances, however, fertilizer recommendations are already expressed in terms of phosphate and potash, so the fertilizer user need not make these calculations.

Most fertilizer labels also include other information, such as percentages of specific nutrient carriers (nitrates, ammonia), chemical formulas, type of release (slow, fast) and type of formulation (powder, granular, liquid), etc.

In turf-grass situations, fertilizers, e.g. urea, ammonium nitrate, super-phosphate, potash and ammonium sulphate, often contain only one of the primary nutrients. Others, such as ammonium phosphate or potassium nitrate, may contain two. Still others are blends of several nutrient carriers. Fertilizers that contain all three primary nutrients – nitrogen, phosphorus and potassium – are called *complete fertilizers*. Other classifications are 'low analysis' fertilizers, in which the three primary nutrients make up less than 30% of the total (e.g. sewage sludge and most organic fertilizers) and 'high analysis' fertilizers, in which the combined total is greater than 30% (e.g. most inorganic fertilizers).

## Fertilizer Sources

Even though 16 different mineral elements are essential for the growth of turf-grass, nitrogen is by far the most important. It has a dramatic impact on turf-grass colour, growth, density, tolerance to stress and recuperative ability.

A wide selection of nitrogen sources is available for use in turf-grass situations. Some are the same for turf and agronomic crops, whereas others are unique to turf-grass. An understanding of the various nitrogen sources is critical in determining the best nitrogen source for a particular turf-grass situation.

In general, nitrogen sources can be separated into the groups presented in Fig. 5.3. Many of them are relatively easy to manufacture and therefore inexpensive. Figure 5.4 presents examples of chemical synthesis of several popular nitrogen fertilizers.

### Inorganic nitrogen carriers

Ammonium sulphate, ammonium nitrate, calcium nitrate and potassium nitrate are common inorganic fertilizers. Once they are applied to the soil,

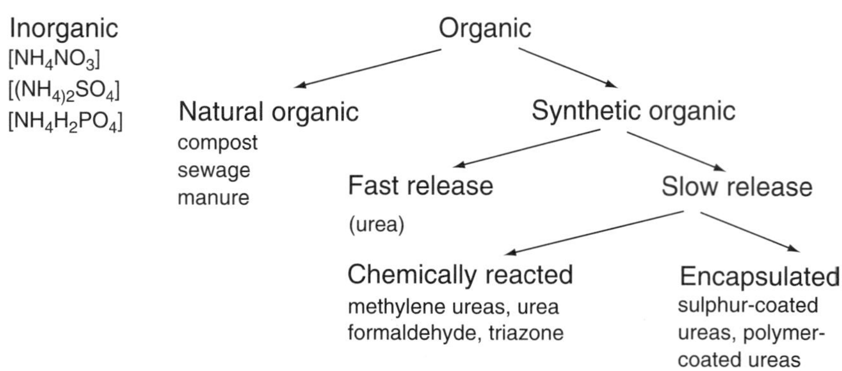


**Fig. 5.3.** Nitrogen sources used in turf-grass management.

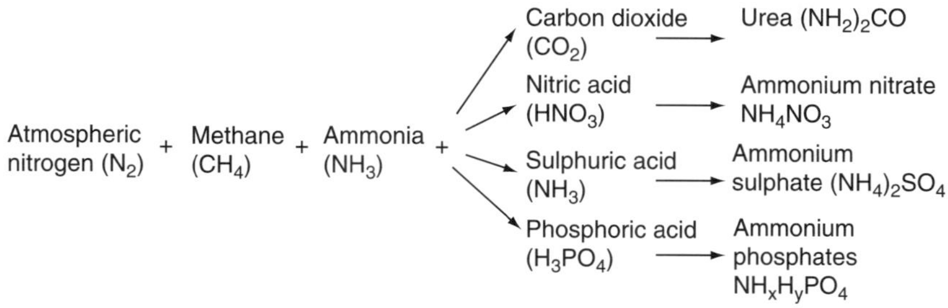


**Fig. 5.4.** Synthesis of nitrogen fertilizers.

they quickly dissolve. Ammonium ions ($NH_4^+$) are held in the soil by the negatively charged clay particles and by organic matter. They can be taken up by the turf either immediately or at a later time. Dissolved nitrate ions ($NO_3^-$) become available immediately for uptake but are not held in the soil. Any that are not taken up at once are subject to loss by leaching.

Inorganic nitrogen carriers are the least expensive, but they are also the most unsafe to apply. Application of these fertilizers in the powder form can cause profound damage to the turf. They are all hygroscopic (i.e. they absorb moisture from the air), and when they come in contact with the leaves they burn them severely. Granular forms are safer because much less of the leaf area is in contact with the granules, so less burning takes place. Light watering immediately after fertilizer application is helpful and highly recommended.

## Organic nitrogen carriers

Organic nitrogen carriers can be divided into two major categories: natural and synthetic. In natural organic fertilizers, nitrogen originates from plant or animal sources. Composts, sewage sludge and similar types of products belong to this group. In synthetic organics, the nitrogen originates from atmospheric air, and fertilizers are mass-produced in the process of organic synthesis. Urea is the most important fertilizer in this group.

Natural organic sources have gained increasing popularity, especially in the home-lawn industry. Much of the public worldwide considers natural products better than synthetic ones and less damaging to the natural environment. In general, natural fertilizers are excellent sources of nitrogen, as well as other nutrients, needed by turf-grass plants and assure their healthy growth. They usually contain less than 10% of slow-release nitrogen and cause very little burning. Sewage-based products dominate organic fertilizer markets. They are safe and easy to use. Municipal sewage is first dried, then heated and finally granulated. The finished product is usually free of pathogenic bacteria and diseases and seldom has an unpleasant odour. Sometimes sewage-based products contain excessive levels of heavy metals, especially if produced in regions with heavy industrialization. Only reputable products from frequently monitored sources should be applied. Animal manure, blood meal, poultry feathers, plant by-products and numerous other fertilizers are produced around the world. Even though nitrogen and other nutrients from natural fertilizers are substantially more expensive than those from synthetic fertilizers, their usage has steadily increased.

### Synthetic organics

Urea is a low-cost nitrogen source for turf-grass and is the most widely used. Commonly used granulated urea is a fast-release, completely soluble product. Granulated urea looks similar to ammonium nitrate and contains even more nitrogen, but as an organic compound it has much less potential to burn turf-grass leaves. After application, it dissolves in the soil water but cannot be taken up by plants before the enzyme urease, which is naturally present in the soil, breaks urea's molecules and releases ammonia. Ammonia reacts with water, forming ammonium ions ($NH_4^+$). Ammonium ions are held by soil colloids and in this form are available for plants. Unfortunately, part of the ammonia can be lost from the soil in its gaseous form through the process of volatilization. Sometimes leaching also occurs before dissolved urea is converted to ammonia ions. Volatilization losses occur especially in sandy soils and leaching under high-rainfall (monsoon) conditions. Even though urea's leaf-burning potential is lower than those of inorganic carriers, it is still a concern. To reduce leaf-burning potential, as well as volatilization and leaching losses, fertilizer researchers have developed numerous ways to slow down the release of nitrogen from the urea.

One major approach is to combine urea chemically with some other relatively inexpensive compound. Another is to encapsulate urea pellets within less permeable material.

Examples of products that have resulted from the first approach are urea formaldehyde (also called UF), triazone and isobutylidine diurea (called IBDU). Examples of encapsulated products are sulphur-coated urea, resin-coated urea and combinations of the two (Fig. 5.5). Most of the urea products described below are used frequently, and the majority of them are commercially available in most countries. Unlike urea or ammonium nitrate, these types of fertilizers are sold under particular trade names, which may differ in different countries. The fertilizer label

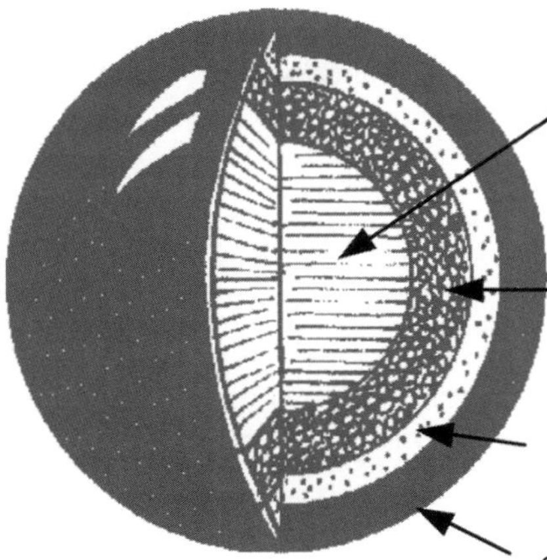


**Fig. 5.5.** A pellet of sulphur-coated urea.

provides specific product information and lists the product's composition, time of release, etc.

Sulphur-coated urea (SCU) is produced when granulated urea is passed through a stream of molten sulphur, with which it becomes coated. A wax coating is often applied in addition to sulphur to protect the surface from microbial degradation. Sulphur-coated urea is a relatively low-cost slow-release nitrogen source. The rate of diffusion of urea from the coated granules depends upon the thickness and integrity of the coating. The thicker the coating, the slower the release will be.

Manufacture of resin (or plastic)-coated urea uses a technology similar to that for sulphur coating. Other soluble nitrogen sources besides urea, such as nitrate and ammonia sources, are also resin coated. Release of nitrogen from resin-coated products depends on osmosis through the semipermeable resin coating rather than on coating imperfections as in sulphur- or polymer-coated products. Plastic-coated urea works on the same principle as the resin-coated product. Both resin- and plastic-coated products have more predictable release characteristics than sulphur-coated products and give turf managers a high degree of control over nitrogen release.

### Methylene urea

The chemical combination of urea and formaldehyde creates urea-formaldehyde compounds commonly referred to as methylene urea fertilizers. These products occur as polymers of varying length. The smaller the ratio of urea to formaldehyde, the longer the chain of polymers

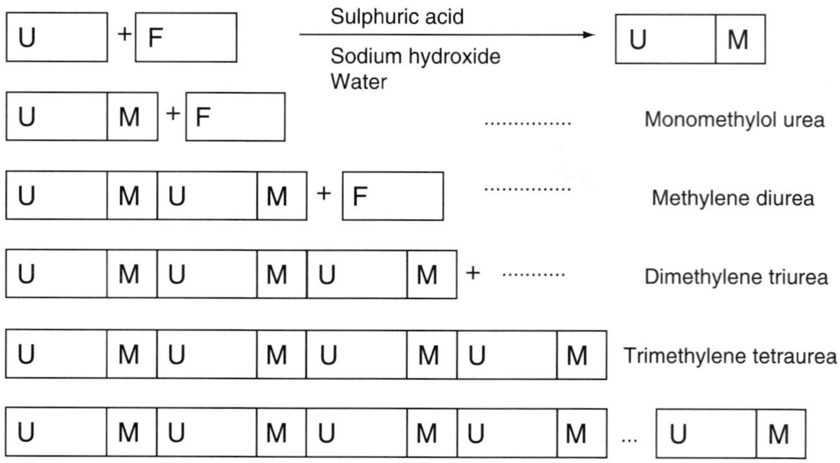


**Fig. 5.6.** The reaction products of urea (U) and formaldehyde (F). M = methyl group.

formed. As the polymer lengthens and the number of 'chain links' increases, solubility decreases and nitrogen is released more slowly. Urea formaldehyde is one of the oldest controlled-release forms of nitrogen. Developed in the 1940s, it is still manufactured and universally used. The polymerization process, presented in Fig. 5.6, starts with the reaction of urea and formaldehyde. The product of this reaction, monomethylol urea, closely resembles urea but releases nitrogen a little more slowly and has somewhat reduced potential for leaf burning. Addition of another urea molecule to the chain produces methylene diurea (called MDU) which has even slower nitrogen release and less burning potential. Addition of the next one produces dimethylene triurea (called DMTU). Then comes trimethylene tetraurea, and so on. Many of these polymers are mixed with other forms of nitrogen and sometimes with other nutrients, organic fertilizers, etc. Long-chain urea formaldehyde polymers are among the slowest-release forms of nitrogen available.

Isobutylidene diurea (IBDU) is made by reaction of isobutylaldehyde and urea. The solubility of this product is very low, but in the presence of soil moisture, IBDU hydrolyses back to urea and butylic acid. The released urea is converted to ammonia and in this form is available to plants. Besides level of moisture, the rate of nitrogen release is governed by the particle size of this product, which is generally available in coarse, fine and powder grades. Because the ratio of surface area to volume of small particles is larger than that of large ones, smaller particles hydrolyse faster. Even in powder form, IBDU releases nitrogen relatively slowly, but if very slow release is desired, coarse grades should be applied. Several other types of nitrogen sources are the subjects of ongoing research. Fertilizer manufacturers continually discover improved sources and bring them to market. Modern turf-grass management with complex fertilization programmes often uses a combination of several nitrogen sources. Manufacturers serve the needs of the turf industry by providing fertilizers that contain various nutrients in both quick-release and slow-release forms.

### Phosphorus carriers

Far fewer sources of phosphorus than of nitrogen are used in turf-grass situations. Phosphorus is usually applied to the soil as superphosphate (20% $P_2O_5$) or triple superphosphate (46% $P_2O_5$). Two other water-soluble ammonium phosphate sources, known as MAP (monoammonium phosphate 12-61-0) and DAP (diammonium phosphate 18-46-0), supply both nitrogen and phosphorus. In addition to easily soluble forms of phosphorus, sulphur-coated or polymer-coated fertilizers are available in some countries, but their importance is marginal.

### Potassium carriers

Two sources of potassium are especially popular, potassium chloride (60% $K_2O$) and potassium sulphate (50% $K_2O$). Potassium sulphate is usually a little more expensive but is less likely to burn foliage. A third popular source is potassium nitrate (13-0-44), which supplies both potassium and nitrogen. Slow-release potassium is available in the form of sulphur- or polymer-coated products.

### Iron carriers

Several sources of iron are applied to the soil or to foliage. On acidic soils (pH below 6.5), ferrous sulphate (20% Fe) or ferrous ammonium sulphate (14% Fe) are used most often. These forms are insoluble at higher pH and under those conditions should be replaced with a so-called chelated form of iron. A chelator is a large organic molecule that holds iron bound to its carbon ring structure. Chemical binding forces are relatively weak but strong enough to prevent iron from rapidly changing into forms that are unavailable to plants. Fertilizers made up of iron chelate contain 7–10% Fe.

### Micronutrients

As previously mentioned, micronutrients are seldom used in turf-grass management. Fertilizer companies rarely manufacture them specifically for turf. If really needed, they can be found at agrochemical supply outlets.

## Methods of Application

Unlike those used on agronomic crops, turf fertilizers must be applied over the foliage of the already growing plants. Both the correct amount of fertilizer and uniform coverage are essential. Dry fertilizers can be applied with either a drop-type (gravity) spreader or a rotary (centrifugal) spreader (Fig. 5.7). A drop-type spreader is often used on relatively small areas or close to flower-beds, shrubs, buildings etc. Its major advantage is its accurate application pattern. Its disadvantages are considerably slower operation and the requirement that each pass meets exactly with the previous one; otherwise, skips or overlaps in the application will be noticeable. The rotary (or centrifugal) spreader has a wider pattern of distribution and can therefore cover a larger area in a shorter time. In addition, the density of its application pattern gradually diminishes away from the machine, reducing the probability of visible skips or overlaps in fertilizer application. Centrifugal spreaders are more difficult to calibrate, and the application is less precise, but in most situations, they are usually

**Fig. 5.7.** Drop-type and centrifugal spreaders are most commonly used in turf-grass management.

sufficiently accurate. Methods of spreader calibration are listed in the Appendix.

Small areas such as home lawns can be fertilized with hand-operated spreaders (Fig. 5.8). The previously determined amount of fertilizer should be placed in the spreader and applied as evenly as possible over the intended turf area (see measurements and calculations in the Appendix). A jar with holes punched in the lid can be used as a simple fertilizer applicator on very small areas. Granulated fertilizer is placed in the jar, which is shaken over the plants so that the fertilizer is scattered through the holes. The jar can be refilled several times if necessary to spread the required amount of fertilizer. The holes should be small enough to prevent overapplication. White pellets on the green turf guide an applicator very efficiently by indicating which spots have received more fertilizer and which have received less. This basic method of fertilizer application usually assures more uniform coverage than hand-operated spreaders.

Liquid fertilizers can be applied to the soil or to the foliage. Commercial lawn-care companies use a soil-drenching technique in which fertilizers dissolved in water are applied with a hose or with special applicators. Foliar feeding, mostly with nitrogen or iron, is used when nutrients must be absorbed directly by the turf-grass leaves. This technique is often used when turf managers want to mask a burn that resulted from herbicides or other pesticide applications or when they wish to improve the colour of the turf before important sporting events. Sometimes pesticide manufacturers add liquid fertilizers to herbicides or other pesticide formulation to mask turf injury resulting from pesticide application.

**Fig. 5.8.** A hand-operated spreader.

## Fertilization Frequency

Theoretically, it would be most beneficial to apply a very small amount of fertilizer every day. Obviously, however, this practice is rarely used because it is impractical and expensive. More feasible is to apply fertilizers frequently enough to maintain consistently adequate and balanced levels of nutrients in the soil. The frequent practice of overloading the soil with nutrients, then depleting it to the level of deficiency, then overloading it again should be avoided. When a fertility programme is planned, especially in a humid climate, an important factor is intensity of rainfall. As already mentioned, the most important nutrient, nitrogen, is the easiest one to lose with rainwater. Several days of intense monsoon may drain virtually all the soluble nitrogen from the root zone. In this situation, the fertilization programme should be adjusted.

When fast-release fertilizers are used under relatively stable weather conditions, high-maintenance turfs such as golf greens should usually be fertilized once a month (sometimes biweekly), medium-maintenance turfs such as fairways bimonthly, and home lawns 3–4 times a year. In the arid or semiarid tropics, intervals will be somewhat longer. In the humid tropics, especially during the rainy season when slow-release products are used, fertilizer should be applied only one-third or one-quarter as often, depending on the type of product used. It is important to remember that, if slow-release fertilizers are used, rapid changes of fertilization programme are not possible. Sometimes a rapid and drastic reduction of nitrogen fertilization is needed, i.e. in response to the sudden occurrence of a certain disease or insect pest (see Chapter 11, this volume). A heavy load of slow-release fertilizer cannot be washed out of the soil for months, so turf managers are often unwilling to use slow-release nitrogen carriers on high-value turfs and prefer instead to apply otherwise less convenient fast-release fertilizers.

# 6 Mowing

Sound mowing is perhaps the single most important factor contributing to the attractiveness and longevity of any turf-grass area. Cutting leaves off at a uniform height produces smooth turf with an attractive appearance. Cutting through stems (stolons, rhizomes and tillers) causes the turf-grass plants to produce more stems, which grow roots of their own to produce more individual plants, and therefore maintains or increases turf density. Turf managers ordinarily use mechanical mowers powered by internal-combustion or electrical engines, but occasionally, especially on small areas, hand-pushed mowers are still used. Before modern times grass was mowed by grazing animals, mostly sheep, which provided a rather rough quality of cut. The first steam and push-type mowers were introduced in the late 1800s (Fig. 6.1). Today, modern mowers provide a high-quality cut, and those used on recreational or sport turfs can be adjusted with great precision, often to within 1 mm of the desired mowing height.

Among the major benefits that result from frequent mowing are improvements of turf-grass appearance and preservation of the plants' health. Properly mowed turf is frequently denser, more resistant to invasion by weeds, and more resistant to traffic damage, diseases, pests and numerous other stresses.

## Turf Responses to Mowing

Mowing is always stressful for turf-grasses. Cutting leaf tissue disrupts physiological processes and creates open wounds in the tissue through

**Fig. 6.1.** Early mower used around 1900.

which unwanted pathogens or other organisms can enter the plant. Removal of leaf area reduces the plant's capacity to carry out photosynthesis and consequently lowers production of carbohydrates. Reduced production of carbohydrates results in decreased production of new roots, which in turn results in decreased abilities to draw water and nutrients from the soil. All these negative factors would seem to indicate that grasses should be mowed as infrequently as possible, and that conclusion is correct for over 10,000 species of wild and forage grasses around the world, but the 40–50 species of the so-called turf-grasses are different. They have the unique ability to compensate for the loss of leaf tissue to mowing by increasing their density below the mowing height. For these species and varieties mowing is still somewhat stressful, but it is not damaging if constant mowing height is maintained. The proper mowing height and frequency are those that provide the optimal balance between desired appearance of the turf and the physiological abilities of turf-grasses to withstand mowing stress. Much scientific research has been conducted to determine this balance not only for each turf-grass species but also for each turf-grass variety. Even occasional mowing below the point where stems branch is tremendously stressful, so removing only a modest portion of the leaves and keeping consistent mowing height are crucial.

## Mowing Height

The height to which a given grass can be mowed is directly related to its ability to produce enough leaves and to keep up with production of carbohydrates. This ability is determined by the growth habit and type of the grass. Factors such as the length of internodes, the number of stolons or rhizomes, the height of the crown above the soil surface and natural vigour all influence the amount of leaf mass produced by a particular grass species and determine its ability to withstand low mowing heights. Some creeping grasses with fine leaves, such as cynodon and zoysia, when properly fertilized and watered, are able to produce adequate leaf surface at very low mowing heights. Species with larger leaves and elevated stolons and crowns, such as St Augustine grass and centipede grass, must be cut higher. As a general rule, turf-grasses with fine leaves can be cut lower than those with larger leaves. Table 6.1 presents preferred mowing heights for turf-grasses commonly found in the tropics. Variation between cultivars can be substantial, so the values given should serve only as guidelines.

Turf should not be cut lower than the height recommended for the particular species or cultivar, but mowing somewhat higher is allowable if the result does not interfere with the intended use of the turf. If a lower mowing height should be desired in the future, however, the height must be lowered gradually. The process must be spread over at least four or five consecutive mowings. Making the change in one step is likely to produce 'scalping' (removal of too much leaf area and too many crowns) and will subject the turf-grass to severe physiological stress.

Mowing height is directly correlated with growth of the root system, and higher mowing promotes both greater total root mass and greater rooting depth (Fig. 6.2). Larger numbers of deep roots increase the turf-grass's ability to draw water from deeper soil zones, and a larger root system overall promotes absorption of nutrients from the most fertile zone, the topsoil. Residential lawns and parks can routinely be mowed at the maximum heights recommended in Table 6.1, or a little above. It is seldom beneficial to mow lower, but mowing much higher should also be avoided, even if consistent with the purpose of the turf area. As mentioned before, a mower's knife cuts through the stems, which stimulates their branching and enhances turf density. If turf is mowed too high, too few stems are cut, so stems branch less and put more energy into vertical growth. As a result, turf density decreases, and the soil surface may

**Table 6.1.** Recommended mowing heights for tropical turf-grass species.

| Species | Mowing height (mm) |
|---|---|
| Cynodon | 5–35 |
| Paspalum | 10–50 |
| Zoysia | 10–50 |
| Carpet grass | 25–75 |
| Centipede grass | 25–75 |
| Bahia grass | 40–80 |
| St Augustine grass | 60–100 |

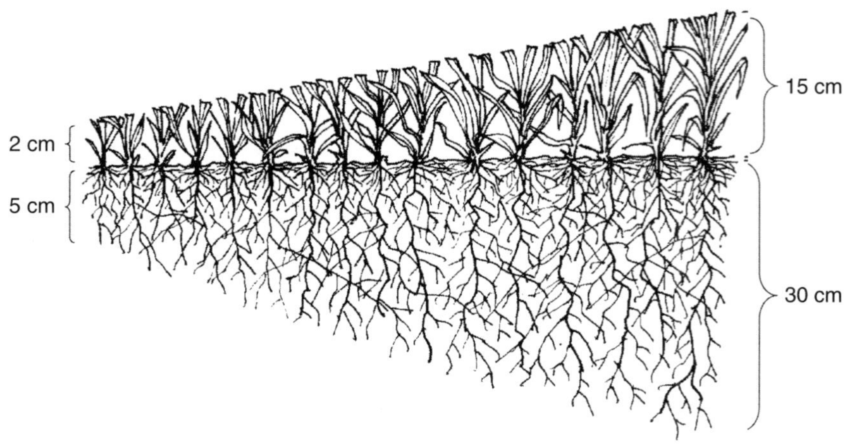


**Fig. 6.2.** Greater mowing height reduces shoot density but favours deep rooting. (Drawing by R. Castro.)

become exposed. Bare soil creates very favourable conditions for germination of weed seeds and may result in severe weed infestation. Mowing height is sometimes imposed by the purpose for which the turf is maintained. Golf greens, tennis courts, bowling greens and similar areas must provide surfaces of particular quality, on which a ball would roll or bounce at a specific speed or in a particular manner. These turfs must be often mowed lower than is desirable from a physiological standpoint. Maintenance of such areas can be extremely challenging and in some cases impossible. As can easily be seen on television screens during transmissions of sporting events, certain areas of sport fields or courts do not hold despite superior-quality turf and superb management. Occasionally the cutting height on a sports turf must be a compromise between the demands of a specific game and the hardiness of the turfgrass. Extensive research aimed at reducing minimal mowing height is being conducted around the world, but its accomplishments usually trail behind demand.

## Mowing Frequency

Mowing too infrequently allows the grass to grow so tall that any subsequent mowing removes too much leaf area. It has been determined that removal of more than one-third of total leaf area results in severe physiological shock to the plant, greatly restricting carbohydrate production and often causing excessive graying or browning of the leaf tips. The plant must use all its carbohydrate reserves to repair the damage, as well as to build leaf tissue in order to restore photosynthesis as quickly as possible. All these resources are used at the expense of the roots, causing a major portion of the young roots to die and interrupting the growth of others. The greater the percentage of leaf removal, the longer root growth is disrupted. In addition to physiological stress, the accumulation of excessive clippings provides favourable conditions for disease development, as well as harbouring unwanted insects.

Unlike mowing height, mowing frequency cannot be specified on the basis of turf-grass variety. Only the growth rate determines mowing frequency. Because no more than one-third of the leave area should be removed at any one mowing (Fig. 6.3), mowing frequency generally increases as mowing height is lowered. For example, if cynodon grows 1 mm per day, a golf green cut to 3 mm should be mowed when its height reaches slightly above 4 mm, that is, daily. The same cynodon on the fairway, maintained at a height of 25 mm, should be mowed when its height reaches 33 mm, that is, weekly. Mowing frequency still cannot be specified in terms of days per millimetre of mowing height, because turf-grass growth rate is influenced by weather conditions, moisture conditions, soil fertility and the natural growth rate of the turf-grass involved. Only the growth of the turf-grass, together with the rule that no more than about one-third of the leaf area should be removed at each mowing, can dictate the proper mowing frequency.

Mowing frequency is sometimes influenced by seed-head formation. Some turf-grasses, such as common cynodon, Bahia grass, carpet grass and St Augustine grass, may produce seed-heads between regularly scheduled mowings. Seed-heads detract from the visual quality of turf and, if unacceptable, must be mowed off. Mowing can also remove flowering parts of certain weeds, before seeds are formed and spread.

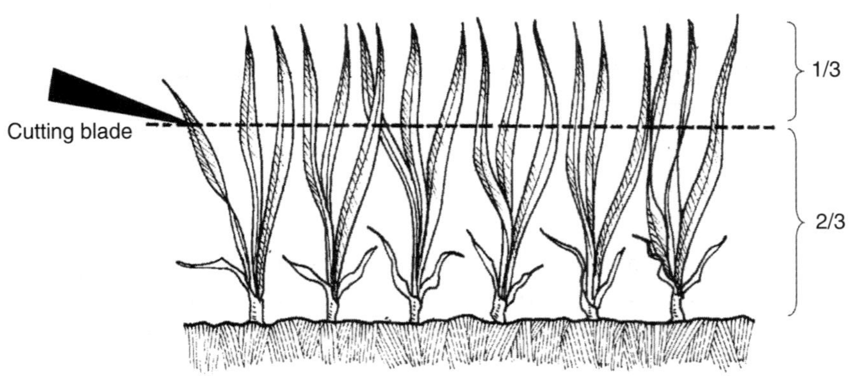


**Fig. 6.3.** No more than one-third of the leaf area should be removed at any one mowing. (Drawing by R. Castro.)

## Scalping

Scalping is the process of mowing turf too short. Scalped turf usually appears brown because healthy leaves have been removed and turf crowns, dead leaves or even the bare soil have been exposed. Incorrect mower settings and mowing on uneven surfaces can cause scalping, especially on hills or where the mower wheels drop into a lower spot (Fig. 6.4). Scalping should be avoided, but if it occurs, the scalped area should be kept well-irrigated and protected from any additional stresses while it recovers.

Scalping is sometimes used deliberately for the purpose of thatch control, usually on turfs of moderate quality, such as home lawns. Thatch (which will be discussed in greater detail in Chapter 8, this volume) is accumulated dead and living plant material (stems, roots and shoots) that develops between the soil surface and the green portion of leaves. Thatch becomes undesirable when excessively thick and should be controlled. Vertical mowers are designed to remove it, but when they are not available or are too expensive, homeowners sometimes scalp the turf instead. Scalping is a poor substitute for vertical mowing, but its use may delay the need for vertical mowing where thatch build-up is modest. Turf-grasses with rhizomes, like cynodon and zoysia grass, can be scalped down nearly to the soil surface, but turf-grasses without rhizomes or with high crowns are sensitive to scalping and may be killed if they are dethatched in this way.

**Fig. 6.4.** Scalping often occurs when mower wheels drop into a lower spot.

## Clipping Removal

Under frequent, normal mowing, clippings should be left on the surface if possible. In tropical climates, grass clippings decompose quickly, returning so many nutrients to the soil that fertilization needs can be reduced by as much as 30%. As a general rule, clippings should be removed only when they interfere with the appearance or purpose of the turf. For example, on areas such as golf course greens or tees clipping should be removed routinely; on home lawns they should almost never be removed. Contrary to common belief, clippings do not increase thatch build-up. Instead, as a source of easy-to-decompose organic matter, they contribute to development of humus in the soil. Infiltration of water is also improved when grass clippings are left in place. Overall returned clippings play a very positive role in turf grass management.

Clippings should be removed, however, when they are present in excessive amounts. For example, in the humid tropics, long-lasting monsoons sometimes disrupt normal mowing frequency. The large quantities of clippings that can result should be removed so that they do not damage the turf by excluding light. For the same reason, clippings should not be left as undispersed clumps on the turf surface. Many modern mowers are equipped with blades that either force clipping into the turf or throw them relatively far, preventing clumping. Alternative methods are raking, dragging a water-hose over the area and other practices that disperse clippings.

## Mowing Patterns

Frequent and close mowing in the same direction or pattern can impart grain, a condition in which the turf-grass leaves and shoots lean in the direction of the cut. On closely mowed turf, such as golf course putting greens, grain can change the path and speed of a rolling ball and is therefore highly undesirable. Varying the mowing pattern presses the turf shoots in different directions with each mowing and tends to make them grow more upright. In addition to reducing grain, altering the mowing pattern on closely mowed turfs reduces the excessive wear that results when the mower is always turned around in the same locations. Mowing pattern is usually not a concern on turfs mowed higher, such as parks or home lawns. These turfs, especially when mowed with rotary mowers, are less likely to develop grain, and even if they do, their use is generally not affected. Grain is sometimes desirable. Sport turf managers use certain mowing patterns to create particular visual effects. Long strips on athletic fields are often mowed in opposing directions, which creates streaked or squared patterns. This method is an alternative to foliar application of chelated iron, mentioned in Chapter 5, for short-term improvement of turf appearance.

## Plant Growth Regulators

Plant growth regulators (PGRs) are naturally produced chemicals that alter growth of the plant by inhibiting or promoting division, elongation and differentiation of cells. In the turf-grass industry, synthetic PGRs have been used since the 1970s to restrain turf-grass flowering (production of seed-heads), as well as to slow down growth and therefore to reduce the need for frequent mowing. PGRs are appreciated in tropical climates, where growth of turf-grasses is vigorous and mowing is needed year-round. The primary problems with most PGRs have been phytotoxicity and inconsistency in response. Physiological and morphological differences between turf-grass species or even varieties, diverse environmental conditions and unpredictable weather patterns result in inconsistent responses by the plants, and make establishment of rates and application recommendations difficult. Besides these imperfections, PGRs are still expensive and require frequent applications (usually 6–8 times a year) and long-term commitment. If an application programme is begun, it should be continued; otherwise, when PGR treatment wears off, tremendous growth surges occur, making mowing even more frequent and burdensome. In addition, if damaged by diseases, insects, scalping or any other type of stress, PGR-treated turf takes much longer to recover.

Progress in PGR development is continuous. Recent products are less phytotoxic and more reliable. Although PGRs will never replace mowing, the 50–70% reduction they provide can save considerable amounts in labour costs.

Like other chemical agents, PGRs must be applied strictly according to the label recommendations. Just like pesticides, if misused, they can contaminate ground water and prove toxic to animals, aquatic wildlife, pollinating insects, etc.

## Mowing Equipment

Several basic types of mowers are available. The most accurate are reel mowers, rotary mowers are intermediate and sickle-bar mowers are the least precise. A fourth type, commonly called vertical mowers, are used to slice turf vertically rather than to remove the tops of the leaf blades and will be discussed in Chapter 8.

The proper choice of mowing equipment depends on the type of grass and the conditions of usage and maintenance. Sharp, properly adjusted reel mowers are recommended for cutting high-quality sport or recreational areas. They provide a clean, even cut and leave the turf looking the most attractive. Rotary mowers should be used where perfect appearance is not necessary and where ease of operation, maintenance and price are of significant concern. Home lawns are predominantly mowed with rotary mowers. Sickle-bar mowers are the simplest and roughest. They are used mainly along roadsides and in other areas where turf appearance is of secondary concern.

Hand-pushed mowers are still available, but most turf mowers are powered by internal-combustion engines. They are usually efficient, strong and the most reliable. Electric mowers are popular in some countries, particularly in large cities. They are fairly effective, especially when a battery with enough power to cut a lawn on a single charge replaces the inconvenient electrical cord. Tractor-pull mowers are used on large areas, such as parks and golf course fairways.

## Reel mowers

A reel mower consists of a horizontal rotating cylinder, the reel, with attached blades and stationary bed knife, which is parallel to the ground (Fig. 6.5). The reel spins around its long axis, and the scissoring action of its blades against the bed knife cuts the grass leaves. The reel usually bears between three and seven blades, and the width of the mower usually ranges from 50 to 80 cm, depending on the model. Reel mowers are powered by engines, pulled by tractor or sometimes pushed by hand. Small single-unit reel mowers cut short grasses well and are used mainly

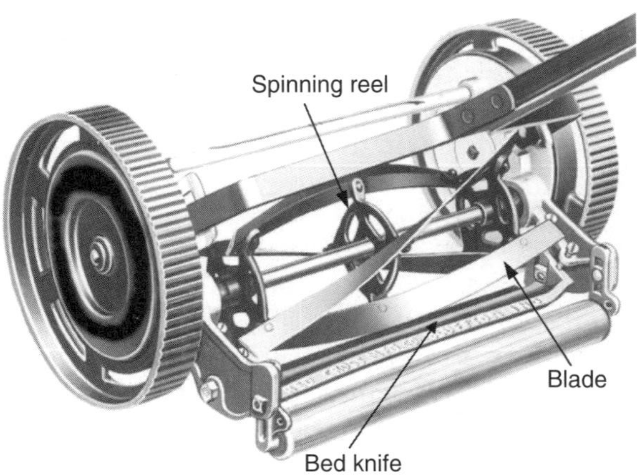


**Fig. 6.5.** A reel mower. The spinning reel sweeps across the bed knife and cuts the grass leaves like scissors.

**Fig. 6.6.** A gang mower pulled by a tractor.

to mow golf course greens and tees and selected sport turfs. Golf course fairways and large parks are often mowed with the so-called gang units of reel mowers, which are pulled by tractors (Fig. 6.6). A gang unit usually consists of 3–9 independent reel mowers and provides the fastest way of mowing large areas. Reel mowers must be operated at a speed that depends on the velocity of the spinning cylinder and the number of blades attached to it. If the speed is too high, the cut becomes uneven. Sharpness of the bed knife and spinning blades is of critical importance. Reel mowers used to cut golf greens are sharpened at least weekly. Despite their high quality of cut, reel mowers are not popular among homeowners. A reliable engine-powered reel mower costs several times as much as a rotary mower, requires considerable maintenance and does not work well if the turf is overgrown, as residential lawns sometimes are.

### Rotary mowers

The blade of a rotary mower is usually a single sharpened metal bar suspended parallel to the ground at its centre point. It spins in a horizontal plane, striking and severing vertically growing leaf blades (Fig. 6.7). Rotary mowers do not provide an even cut, and they cause a certain amount of mutilation to the leaf blade at the point of impact. For the majority of turf-grasses rotary mowers provide a satisfactory cut, but some turf-grasses with rigid leaves, such as zoysia, may sustain considerable damage to the leaf tips, especially when the blade is not quite sharp (Fig. 6.8). In general, rotary mowers are designed to mow grass taller than 3 cm, so their operation is restricted to medium- and low-quality turfs, and their primary users are homeowners. Rotary mowers, usually powered by small engines, can cut a wide variety of grasses, require little maintenance and need only periodical sharpening. The shape of the mower blade can be varied to affect the mower's function and the way grass clippings are disposed of. The spinning blade creates air movement under the mower housing that may blow clippings to the

**Fig. 6.7.** The blade of a rotary mower spins in a horizontal plane, striking and severing vertically growing leaf blades.

side, into a bag, or back into the turf. In recent years, the last type, called mulching mowers, have become very popular. The outer section of the mower blade pulls growing turf-grass blades upward to a partial vacuum created by the high-speed rotations and cuts them. The air currents inside the housing keep the clippings swirling around and chop them into fine pieces. Finally, the blade's inner curve directs air downward to force the small clippings into the turf, where they are not noticeable and can decompose.

The flail mower is another kind of rotary mower, in which the metal blade is replaced by several nylon strings rotating in the same manner but with less risk to the operator. The most common injuries inflicted by rotary mowers are the result of rocks or other objects thrown by the blades or consist of cuts to the toes. Nylon strings eliminate many injuries and improve operator safety, but heavy leather shoes should nevertheless always be worn during mowing.

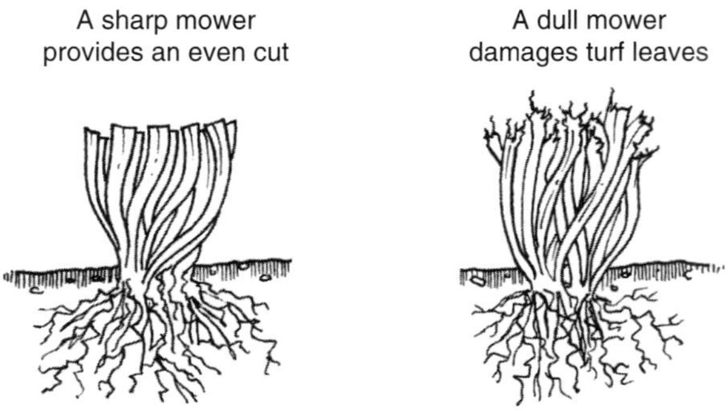


**Fig. 6.8.** Rigid-leaf turf-grasses may sustain considerable damage when the mower blade is dull. (Drawing by R. Castro.)

## Sickle-bar mowers

A sickle-bar mower consists of a multisectional knife that moves in a reciprocating action across plates (Fig. 6.9). It is designed to cut long grass, tall weeds and even light scrub along roadsides and other low-maintenance turfs. Sickle bar units are usually from 1 to 2 m long, often mounted on tractors and used primarily on areas where mowing is very infrequent and quality of cut unimportant. Besides roadsides, sickle bars are best adopted for use on steep slopes and to mow ditches.

**Fig. 6.9.** A sickle-bar mower, used for cutting long grass and tall weeds on low-maintenance turfs.

# 7 Irrigation

Precipitation in the tropics ranges from just about zero to over 2000 mm per year. Warm-season turf-grasses are able to use from 2 to 5 mm of water per day, depending upon location, species, weather conditions, type of maintenance and several other factors. Unfortunately the highest usage of water occurs in arid regions, where rainfall is small, and the lowest occurs in the humid tropics, where rainfall is highest. Even though total yearly precipitation in many regions is more than adequate, its seasonal distribution is usually unsatisfactory to maintain a dense, green turf of high quality year-round. Except in a few localities in the tropics, medium- and high-quality turfs require supplemental irrigation.

## Irrigation Needs

The water required by growing turf may originate from rainfall, irrigation, or a combination of the two. By far the most important factor that determines the plant's need for water is *evapotranspiration*. This term combines the words *evaporation*, meaning water loss from the surface, and *transpiration*, meaning water loss from the plant. It refers to total loss of water from soil covered by vegetation. In turf, where the soil is usually completely covered by growing leaves and stems, most of the water loss is due to transpiration.

Water enters turf-grass through the roots and exits through small openings in the leaf cuticle called *stomata*. Stomata can open and close in response to changing environmental conditions and therefore can regulate water loss from the plant. Many factors influence evapotranspiration, but the most important are humidity, temperature, wind and canopy resistance (Fig. 7.1).

## Humidity

Water loss from the plant occurs because of the gradient that exists between the water-saturated cells and the moisture in the surrounding air. In drier air, the gradient is steeper and water is therefore lost faster. As would be expected, the highest evapotranspiration rates are in arid climates, where the humidity is the lowest.

## Temperature

Increasing temperature directly accelerates evaporation, but temperature's effect on transpiration is more complex. Initially, leaves transpire more in order to cool themselves down, but when temperature rises too high, partial closure of the stomata can be triggered in order to conserve water. Turf-grasses adapted to the tropics use this type of physiological defence less than do cool-season turf-grasses, and unless the temperature is very high, they increase their evapotranspiration when the temperature rises.

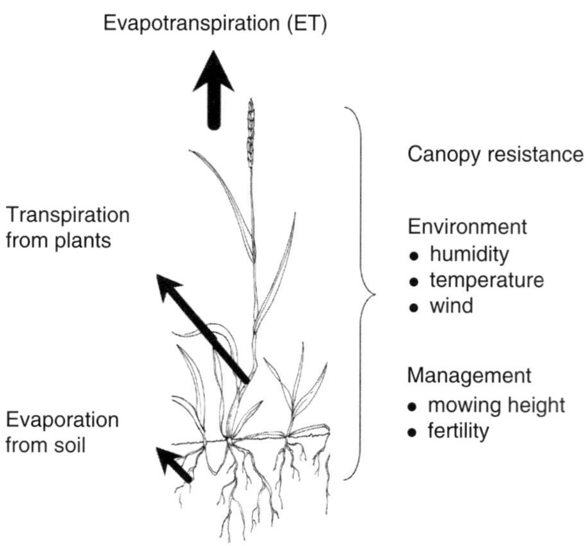


**Fig. 7.1.** Factors influencing evapotranspiration of turf.

## Wind

Movement of air above the turf canopy is an important factor influencing evapotranspiration. In the absence of wind, the leaf surface is surrounded by a thick *boundary layer* of air molecules that block free movement of water molecules diffusing through leaf stomata and thus reduce water loss. Wind disrupts this boundary layer, increasing evapotranspiration, particularly under dry, warm conditions (Fig. 7.2).

## Canopy resistance

Shoot density, leaf orientation, leaf area and growth rate all affect water loss through turf canopy. These factors, described as *canopy resistance*, may either increase or decrease evapotranspiration.

In modern turf management the use of evapotranspiration information is gradually replacing older methods, such as examining soil moisture conditions or visual symptoms of turf. The more advanced irrigation systems use mini weather stations measuring rainfall, air temperature, humidity and wind speed. This information is transmitted to a computer programmed to estimate evapotranspiration for the turf-grass and maintenance regime in question. The computer, rather than a human, decides when the irrigation system should be turned on and how long should it run. Although computer programs are quite efficient and helpful and in the future will probably be used to schedule irrigation systems on small areas such as home lawns, they will never fully replace human knowledge combined with experience and ordinary common sense. A basic

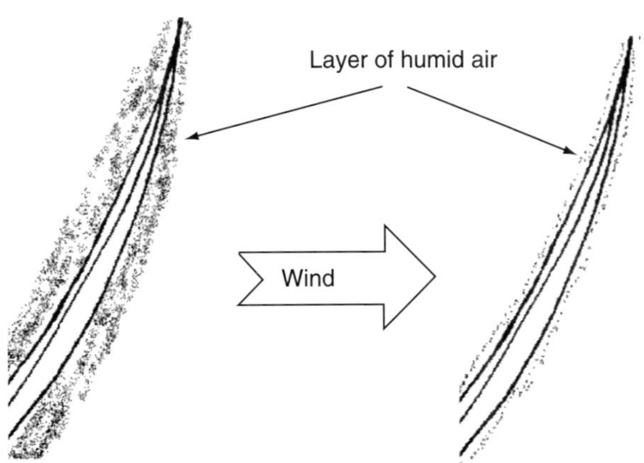


**Fig. 7.2.** Wind disperses the thick boundary layer of humid air surrounding turf-grass leaves.

knowledge of the methods needed for reaching correct decisions is therefore important.

## The Time to Irrigate

Water stress occurs when the rate of water loss through evapotranspiration exceeds the rate of absorption through the root system. Therefore, from the physiological standpoint, the best time to irrigate is just before the turf experiences water stress, which is manifested by wilting. In reality this timing is difficult to achieve because the use patterns of some turf-grass areas may prevent irrigation at the desired time and also because of other issues related to 'turf health' that will be discussed below. Because irrigation systems cannot be turned on, e.g. during football matches or when golfers are playing on the course, turf is usually irrigated only as near to the ideal time as is practical. Irrigation can usually be somewhat delayed. The only critical demand is that it be applied before permanent wilting in order to avoid serious injury and long-lasting damage to the turf.

The simplest visual evidence that wilt is imminent is visibility of prints on the turf canopy. The *foot-printing* technique involves walking across the area and observing how long the turf-grass leaves take to return to their original, upright position. Depending on the turf species, turgid leaves return within one to several minutes, whereas wilted leaves take 15–20 min or more. The foot-printing technique is not always conclusive for species with stiffer leaves, such as zoysia grass or cynodon. An additional guide is turf colour. Water-stressed patches of turf turn bluish green and can be easily distinguished from areas that have not yet undergone water stress. Another method is observation of turf-grass leaves. Water-stressed leaves roll or fold to conserve moisture. When a considerable proportion of leaves are rolled or folded, turf should be irrigated as soon as possible. Examination of soil is also quite reliable. If soil sampled to a depth of 15 cm feels dry, the turf is probably experiencing water stress.

As mentioned above, turf is seldom irrigated exactly at the time just prior to water stress. In addition to factors associated with use of the turf, a major factor that influences irrigation timing is the need to restrain turf disease. Many fungal pathogens require the continuous presence of water droplets for 14–16 h for spore germination and the penetration of turf-grass leaf tissue. Frequent irrigation or irrigation at times that permit water droplets to remain on the leaf for more than 14 h results in increased disease problems (see Chapter 11, this volume). Turf should therefore not be irrigated in the evening or late afternoon. Midday irrigation would be the best if minimizing disease development were the only concern. Unfortunately midday irrigation usually results in substantial water losses due to low midday air humidity and increased evaporation from warmer surfaces and in high drift because the wind is strongest at midday. Early-morning irrigation, around sunrise, is therefore better. Water is applied on the coolest surface and sprayed into the most humid and calmest air. In addition, irrigation water washes off the dew droplets persisting on the leaves, causing turf to dry faster, thus minimizing disease development.

Several other factors can be considered when irrigation is scheduled. Because compaction from foot traffic increases substantially when soil is wet, and running athletes rupture wet turf more easily, many sport fields should not be irrigated within a day or two before a game. During games, sport surfaces should be dry, but at the same time, turf should not be water stressed. In other cases, irrigation scheduling is determined by water availability. Many municipal areas allow watering only at certain times of day or on certain days of the week. Usually, under temporary water-conservation measures, turf is able to survive until restrictions are lifted or eased, but if restrictions are too severe or water extremely expensive, turf may be replaced with the so-called 'alternative landscaping', which includes rocks and drought-resistant plants.

## Irrigation Frequency

The first principle of irrigation management says: 'deep and infrequent'. At each irrigation event, soil capillaries in the entire root zone should be filled with water to soil field capacity and then gradually depleted to the point at which turf-grass approaches light water stress. Excessive irrigation results in water loss, whereas excessive irrigation frequency results in development of shallow root systems (Fig. 7.3). Soil usually dries from the surface, remaining moist at greater depth. Roots seek water where it can be found and therefore often elongate to depths of 50 cm or even more. Turf-grasses with deep roots are more resistant to water stress and perform better overall. Considerable evidence also relates formation of deep roots to a plant hormone called abscisic acid (ABA). Under dry conditions higher production of ABA slows shoot growth and allows more carbohydrates to be translocated to the roots, which induce root growth.

The majority of soils in tropical climates must be irrigated with 10–15 mm of water every 3–4 days. Sandy soils have lower water-holding capacity and may need lighter irrigation daily, especially in arid climates. Heavy clays may have poor infiltration rates, which can result in run-off. Therefore, heavy soils, like sands, may need lighter and more frequent watering.

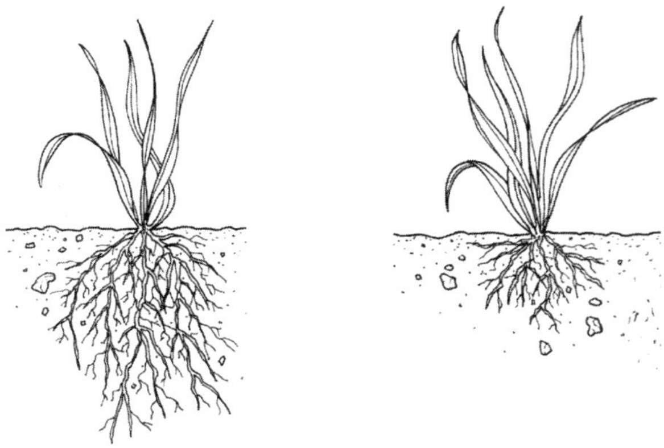


**Fig. 7.3.** Excessively frequent irrigation results in a shallow root system. (Drawing by R. Castro.)

Turfs with excessive thatch layers present unique irrigation problems. Decaying stems and roots in the thatch layer hold irrigation water and reduce its infiltration into the soil. Under dry conditions thatch dries quickly, producing sudden and often severe water stress. Excessively thatched turfs must be irrigated frequently. The 'deep and infrequent' rule must also be modified on newly seeded or sodded areas. In some instances, especially in the dry tropics, irrigation of newly seeded turf may be needed several times a day. One more irrigation technique deserves mention, even though it is seldom used on warm-season species. Turfs in temperate climates dominated by cool-season species occasionally suffer from midday heat stress. When summer weather is exceptionally hot, root systems of certain turfs may not be able to keep up with transpiration demand and may become overheated to the point of death. *Syringing* is the application of small amounts of water for the purpose of cooling, reducing transpiration or preventing turf wilt. Syringing is often performed several times a day.

## Water Sources

The majority of home lawns, parks, landscapes around businesses, sport turfs and other relatively small turf areas are irrigated with municipal water. Golf courses, resorts, large parks and other similar areas generally strive for independent water sources. Connecting to municipal water is the easiest and does not require investments in pumping stations or water-distribution systems, but the water is controlled by municipal government and may be restricted at the time of the highest demand. In times of drought, delivery of water to households for human consumption will always have priority over turfs and landscapes, so access to an independent water source is highly advantageous. The most common independent sources of water for irrigation are wells, lakes, reservoirs, ponds, streams and rivers (Fig. 7.4). Although access to streams and rivers is sometimes regulated by local governments, ponds or lakes, if located on property controlled by the turf manager, provide the most independent water source. Artificial ponds should be constructed small and deep rather than large and shallow and should hold enough water for an extended period of drought. They should be located at the lowest point on the property and recharged from surface drainage or possibly some other source such as natural springs.

Wells provide an excellent water source. The initial cost of developing a well for a large turf area is quite high. Deep wells require large, high-quality pumps and ongoing maintenance to be dependable, but the investment usually pays off quickly. Note, however, that some governments, especially in desert areas, restrict the amount of water that can be pumped out or may restrict the depth of the well.

Streams and rivers are good sources of water if access is unrestricted. Minimum annual flow must be determined, and the flow must exceed the maximum amount of water that is required for irrigation. Besides average minimum flow, at least a 10-year history should be examined, because in dry climates streams and rivers occasionally dry out. If such a danger exists, other water sources should be secured for irrigation of at least the most valuable turfs, such as golf greens and tees.

**Fig. 7.4.** Ponds on golf courses or in parks can store a considerable amount of water.

Finally, municipal effluent discharged from sewage-treatment plants is actually quite a dependable source of irrigation water. Usage of household water does not fluctuate greatly during the year, so it remains available during drought or other periods when potable water may be restricted. Municipal effluent used for irrigation is usually free from pathogenic bacteria, heavy metals and other compounds posing a hazard to humans and animals, but it may contain elevated amounts of certain elements such as sodium and boron which can have detrimental effects on turf. Irrigation with effluent is becoming a global issue. It will be presented more comprehensively later in this chapter.

## Water Quality

Municipal potable water usually presents no problems, but water from other sources may contain constituents that are of minor or major concern.

Solids such as sand, silt and clay particles may sometimes clog water-delivery systems but are generally of little concern. On the other hand, salts dissolved in the water are of major importance, and if their concentration is too high the water may not be suitable for irrigation. High concentration of salts affects the soil-solution osmotic potential, may reduce water uptake by plants, may induce mineral imbalances and, if high enough, can cause direct injury to roots. Total salt content of water is usually measured in terms of electrical conductivity ($EC_w$) and is often reported as decisiemens per metre (dS/m). Salinity hazard is considered low when $EC_w$ is below 0.75 dS/m and very high when $EC_w$ exceeds 3 dS/m.

Of all salts that may be present in water, those of sodium are the most important. High sodium causes soil to disperse, usually resulting in reduced water infiltration and reduced aeration. Sodium ions may also be directly toxic to the roots. If other than municipal water is used, its *sodium absorption ratio* (SAR) should be determined. The same laboratories that conduct soil tests usually conduct water tests. SAR values are reported in milliequivalents per litre (meq/l) and should be below 5 if the water is used to irrigate soils rich in clay, below 10 for use on sandy loams and below 15 for use on predominantly sandy soils. The range is broad because the detrimental effect of sodium is proportional to development of soil structure. Clay soils owe their favourable properties to their granular structure. As was described in Chapter 3, organic and inorganic colloids glue fine soil particles together, creating larger clusters that hold soil water, assure proper aeration, reduce physical impedance and overall provide favourable conditions for root growth. These clusters can easily be damaged and fall apart if sodium ions attach themselves to the cation exchange sites, displacing calcium or magnesium. Sands, on the other hand, do not owe their properties to aggregation of fine soil particles, so presence of sodium ions in sand is not as damaging. If SAR values exceed desired limits, addition of gypsum may relieve the problem. Water-testing laboratories often provide a value called *residual sodium carbonate* (RSC), which indicates how much gypsum should be added to the water to offset the sodium hazard. In some countries, this value is calculated in kilograms of gypsum per hectare of a pond at a standard depth of 1 or 2 m. In others, the calculations are in kilograms of gypsum per 1000 $m^3$ of water.

## Irrigation Methods

Turf is irrigated by three main methods: (i) overhead; (ii) surface; and (iii) subsurface. Overhead irrigation is used on the overwhelming majority of turfs. Water is distributed through some type of irrigation system, either pipes or garden hoses, and sprayed by a sprinkler head. The purpose of a sprinkler head is to disperse water into fine droplets that fall uniformly on the turf surface, as would a light rain. A great variety of sprinkler heads is available on the market. Some are designed for use with high-pressure water lines for irrigating large areas, some for intermediate areas such as parks, some for small areas such as home lawns, some for athletic fields, and so on. They vary greatly in size, design, efficiency, methods of spraying water and material used for their assembly. Except for a few unique types, most can be classified as either rotary or fixed.

### Overhead irrigation

*Rotary sprinkler heads*

Rotary sprinkler heads shoot water as one or more streams of spray. Water flowing through the sprinkler head makes it rotate to cover a circular area or a set portion of a circle (Fig. 7.5).

During sprinkler head rotation, the area of the circle of turf near the sprinkler is much smaller than that of the outer band far from it. A stream of spray from which the same amount of water falls at every distance therefore delivers much more water per unit area to the inner area than to areas farther away. The result is a wedge-shaped irrigation pattern, a direct result of the changing ratio of volume of water to amount of area covered by this water (Fig. 7.6). When two adjacent heads are properly spaced, their overlapping pattern can provide relatively uniform coverage along a line drawn between them. Unfortunately, sprinkler heads are frequently placed in the centres of golf course fairways and sport fields. This type of

**Fig. 7.5.** Rotary sprinkler shooting two different streams of water spray.

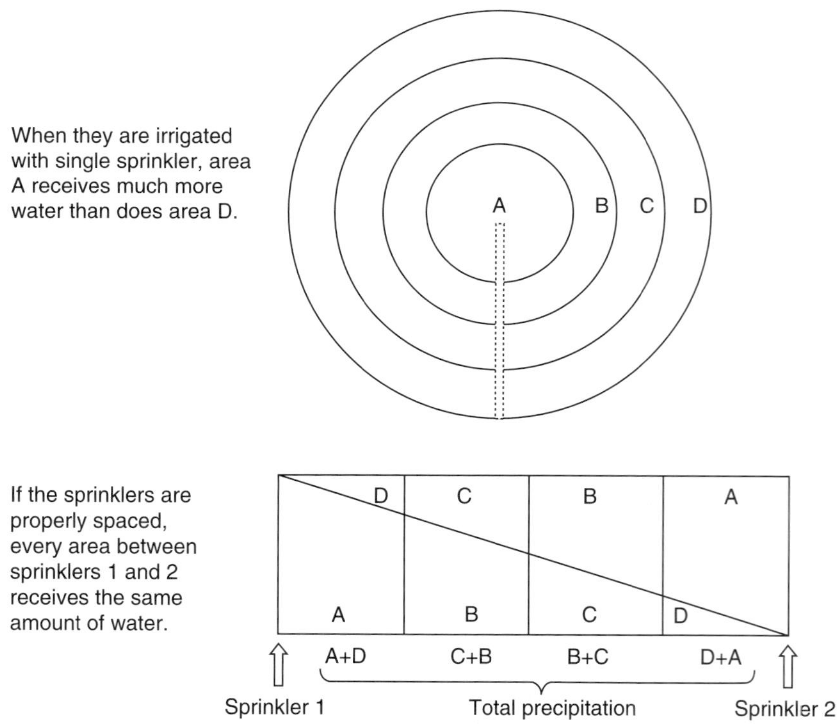


**Fig. 7.6.** The precipitation rate from a rotating sprinkler declines along the radius because the turf area to be covered by same amount of water increases. To assure uniform distribution of water, sprinklers should be properly spaced with their irrigation patterns overlapping.

installation is incorrect and should be avoided. As can be seen in Fig. 7.6, to supply the minimum irrigation needed by the edges of a turf area, a central sprinkler head must substantially overirrigate the central part of the turf. On sandy soil, this overwatering can cause rapid water percolation that results in substantial nutrient leaching, and on heavier soil, it may result in increased soil compaction from foot and vehicular traffic. In all cases, overirrigated areas are more susceptible to diseases, and their soil has less favourable physical properties, which negatively influence turf growth and use of the turf area. Because the single row of sprinkler heads does not assure uniformity of irrigation, the sprinkler heads should instead be placed to cover overlapping half circles from opposite edges of the turf area (Fig. 7.7). The savings associated with installation of half as many underground water lines and sprinkler heads often tempts managers into incorrect decisions with lasting negative consequences.

Rotating sprinklers produce the desired wedge-shaped pattern of irrigation when they are operated within the proper range of water pressure. Excessive high pressure causes the water to form a fine spray instead of droplets. Wind can then divert a large portion of it and cause irregular and unpredictable irrigation patterns. Insufficient water pressure results in streams of water that do not disperse sufficiently and irrigate only a narrow band at some distance from the sprinkler head. When water pressure fluctuates, irrigation can be excessive in some areas and insufficient in others (Fig. 7.8). Wind strongly influences irrigation patterns and water efficiency; often more than do water pressure, sprinkler-head design or sprinkler spacing. Its influence should be minimized as much as possible. The largest sprinkler heads should operate around sunrise, when wind velocity is usually lowest.

**Fig. 7.7.** Sprinklers on the edge overlap with sprinklers in the centre.

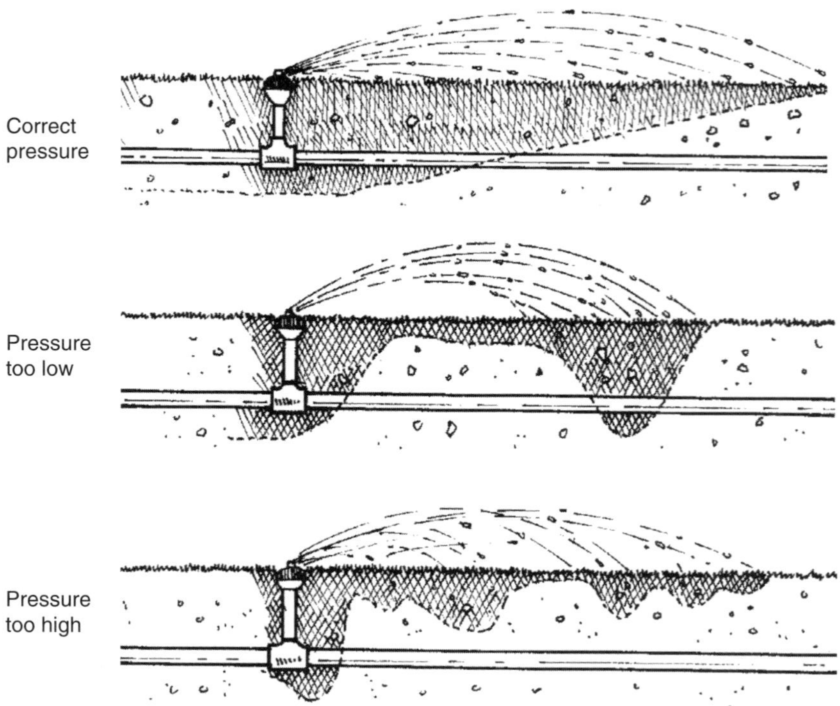


**Fig. 7.8.** To maintain an accurate irrigation pattern, sprinklers must operate under the correct pressure, as indicated by the manufacturer. (Drawing by R. Castro.)

Rotary sprinkler heads are usually available as either pop-up or above-ground systems. A pop-up system is one in which sprinkler heads remain below ground, their flat tops flush with ground level, when not in operation. A group of pop-up sprinkler heads is usually activated by one manual or automated valve. When the value is opened, water pressure pushes the heads above ground. When irrigation is completed and the valve is closed, the sprinkler heads drop back into place, where they are inconspicuous and do not interfere with recreational activities, mowing and other operations. Above-ground systems are positioned completely above the turf surface usually on a mobile or portable piping system. Above-ground systems are used on sod farms and sometimes during turf establishment.

*Fixed sprinkler heads*

Fixed sprinkler heads have no moving parts and usually operate at low water pressure. Each produces a fine spray of water covering a relatively small area, commonly 5–10 m in diameter (Fig. 7.9). They are often used

**Fig. 7.9.** Fixed sprinkler heads usually operate at low water pressure, producing a fine spray covering a small area.

to water home lawns, flowerbeds and other landscape plants. Some sprinkler heads can be adjusted to control the amount of water delivered. They can also be adjusted to deliver water in a variety of spatial patterns. A full head delivers water in a full circular pattern; other typical heads irrigate half and quarter circles. Adjustable heads can water any part of a circle, from 0° to about 360°. Some fixed sprinkler heads are designed to water rectangular and square areas of turf, and end, centre and side-strip patterns are available for grassy pathways, side yards and other tight spaces. Fixed sprinkler heads are most frequently installed as pop-up systems but are sometimes installed above ground. They are relatively trouble free and are the least affected by wind.

*Rain guns*

A rain gun is usually a huge impact-type sprinkler used to irrigate a large turf area. On some sport fields, especially polo fields and horse-racing tracks, they are at times attached to fire hydrants and used for quick irrigation of turf. Their role is rather minor.

*Oscillating sprinklers*

Oscillating sprinklers, also called wave sprinklers, are most commonly used by homeowners on relatively small areas. Water from a garden hose is delivered through a row of small nozzles (holes) in a bent pipe. The pipe turns slowly back and forth, sweeping the row of small streams of water back and forth across the area to be watered.

### Subsurface and surface irrigation

In subsurface irrigation, water is supplied from beneath the soil surface directly to the roots. Perforated plastic tubes called drip lines are most commonly used in this type of irrigation. Such systems make very efficient use of water, but over a period of time the small openings in the drip lines tend to clog with debris, calcium deposits or slime. Subsurface irrigation is most often used on sod farms, where turf-grass is stripped several times a year and drip lines can easily be serviced or replaced.

In surface irrigation, water usually originates from irrigation ditches and flows across the turf-grass area as a result of gravity and the natural slope. It is advisable only for small areas where water is plentiful and inexpensive.

## Irrigation Systems

### Portable irrigation systems

Portable irrigation systems are those that can be moved from one turf area to another. They are used on many turfs, especially relatively small ones, but cannot be considered as efficient as installed systems. Portable systems usually consist of above-ground sprinkler heads attached to flexible and portable hoses.

### Installed irrigation systems

Installed irrigation systems are those in which the system that delivers water to the sprinkler heads is fixed in place, usually underground. Besides sprinkler heads, they include pipes, control systems, valves and, if the system is independent, water pumps. Piping is the basis of a turf-grass irrigation system. Pipes transport water from the water source to the sprinkler head where it is dispersed onto the turf. Proper function of an irrigation system therefore depends on the type, size and condition of pipes. Pipes can be metal, but today nearly all pipes used for irrigation are plastic. They are sufficiently strong, lightweight, relatively inexpensive and resistant to corrosion and rust.

A control system coordinates operation of the entire irrigation system. In manual systems, someone must turn the control valves by hand. An automated system typically includes controllers, which integrate a clock, a timer, and a series of terminals called stations. A programmable controller allows the turf manager to specify the times and locations at which turf will be irrigated. A signal from the controller activates each control valve at the programmed time. The valves, positioned sometimes at individual heads but more often along pipes serving several heads, control the flow of water from the valve to the last sprinkler head in this series. Some automatic systems are equipped with soil-moisture sensors that can override the controller program and prevent the system from being turned on during or shortly after rain, when soil remains moist. Modern, fully computerized, automatic control systems analyse essential weather and soil-moisture data, account for all programmed variations and modifications, and open selected valves at the best times and for the proper periods.

In addition to the remote-control valves used in automatic systems, several other types of valves are used in irrigation systems. A master valve is used to turn off the entire system, check valves are used to limit water flow to one direction, antisyphon valves are used to protect domestic water supplies from back-flow when irrigation is completed, drain valves are used to allow drainage from lines after irrigation, and pressure-regulating valves are used to prevent excessive water pressure.

Larger irrigation systems usually include water pumps (Fig. 7.10). Two types of pumps are used: booster pumps and system-supply pumps. Booster pumps raise in-line water pressure without affecting flow rate. In large irrigation systems, such as on golf courses, booster pumps are used where water pressure drops considerably as a result of elevation changes and must be restored. System-supply pumps are used to draw water from a water source, such as a well, pond or river. A system usually includes several of them, and each supplies a specific flow rate at a

**Fig. 7.10.** Water pumps are usually placed in specially constructed buildings that protect them from weather damage.

specific pressure. When pumps are arranged in series, nominal water pressure is equal to the sum of the capacities of the individual pumps; flow rate remains equal to that of a single pump. When pumps are arranged in parallel, water pressure remains equal to that of a single pump, but flow rate equals the sum of the capacities of the individual pumps (Fig. 7.11).

A well-designed sprinkler system should deliver water evenly to all areas. Sprinkler heads must also be selected and adjusted so they do not spray water beyond turf areas onto streets, buildings or people using parks or playgrounds. Sprinklers should also not be permitted to spray directly onto the trunks of trees or into shrubs and flowerbeds. The sheer force of the water pressure can score tree bark, and perpetual wetting may weaken it, making it more susceptible to pests and diseases. Directed water spray can also blast leaves and flowers and should be avoided.

Installation of an automated sprinkler system requires considerable investment, careful planning, ground excavation, trenching, etc. If the turf area is larger than a small home lawn, the design and groundwork should be done by experienced professionals. Selecting the correct sprinklers, pipes, valves, pumps and control systems requires a competent irrigation designer. Installation of the irrigation system should also be left to a professional who understands the capabilities and limitations of all the system's components. Errors in design or installation result in poor operation, water leaks, damage to sprinkler heads during mowing, etc. and are costly to correct.

## Wetting Agents

Sometimes water applied during irrigation cannot readily penetrate and wet the soil. Wetting agents (surfactants, surface active agents) are compounds able to change water-repellent soils and other surfaces and make

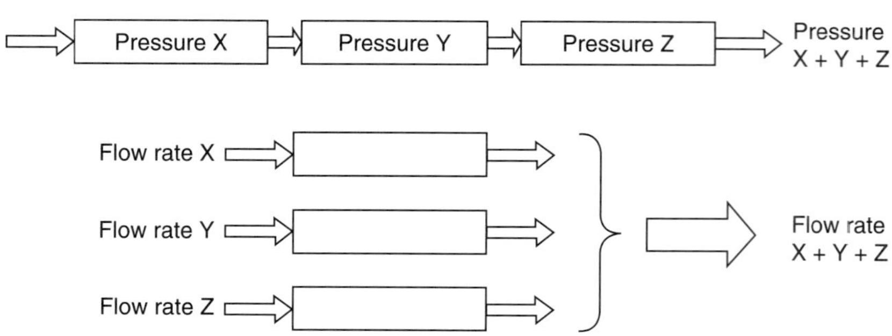


**Fig. 7.11.** Arrangement of water pumps in series influences water pressure, whereas arrangement of pumps in parallel influences the rate of water flow.

them wettable. The mode of action of these compounds is complex, and understanding it requires analysis of forces associated with the behaviour of water molecules. Besides gravity, forces of cohesion, which attract water molecules to each other, and forces of adhesion, which attract water molecules to other substances, play deciding roles in which surfaces can and which cannot be wetted. The effects of these forces can be illustrated by placement of a drop of water on a piece of filter paper and another drop on a piece of waxed paper. On the filter paper, the force of adhesion between the water molecules and the paper molecules is greater than the force of cohesion that holds the water molecules together. As a result, the water droplet spreads out and soaks into the paper. On the waxed paper, the water droplet remains on the surface because water molecules cohere more strongly to each other than to the wax. In this case adhesive forces between water molecules and the surface are weaker than the cohesive forces between water molecules. A surface with these characteristics repels water and is called hydrophobic. Hydrophobic characteristics are found in many types of soils covered by turf-grass, and appearance of hydrophobic localized dry spots, especially on golf course greens and other areas mowed low, is quite common. In numerous studies, microscopic analyses of soil particles taken from these spots have revealed that they are coated with substances that repel water, much like a wax. The coating material appeared to be a complex organic compound produced by a fungus. The localized dry spots become a serious turf-management problem, especially during periods of drought. Despite frequent irrigation, the soil in these spots resists wetting, resulting in patches of dead or severely wilted turf. The water applied wets the turf but does not adequately penetrate the soil surface to reach the root zone.

The use of wetting agents on water-repellent soils can improve their ability to absorb water and thus improve the appearance of the turf (Fig. 7.12). Long-term residual surfactants actively reduce forces of water cohesion in the soil for a period of 6–12 months and should be

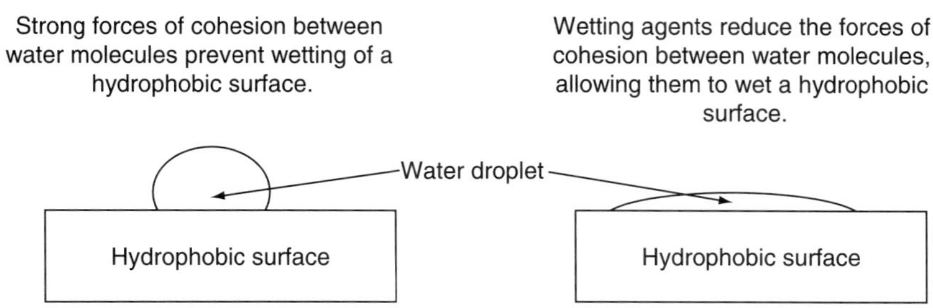


**Fig. 7.12.** Wetting agents reduce the effect of soil hydrophobicity.

reapplied before hydrophobicity returns and turf sustains severe damage. Unfortunately the problem of localized dry spots is chronic, and once it appears in a specific location, it usually persists for many years.

Wetting agents are also effective in irrigation of turfs with substantial thatch build-up. Thatch is often difficult to wet and under regular irrigation can result in increased surface run-off and poor soil infiltration. Addition of wetting agents improves water penetration through the thatch, making irrigation and wetting from precipitation more efficient and uniform.

Wetting agents are also included in the formulation of pesticides. Their influence on efficacy of pesticides will be discussed in Chapter 11.

# 8

# Turf Cultivation, Compaction and Thatch

In agriculture, the term *cultivation* refers mainly to the tilling of soil, but in turf-grass management, it refers to the mechanical processes used to loosen the soil, break up undesired soil layers, remove thatch, stimulate turf growth, as well as to modify the soil surface. Cultivation is usually conducted on higher-quality turfs and involves techniques such as coring, slicing, spiking, forking, water injection, verticutting, top-dressing and rolling. Cultivation, although necessary, also causes a certain amount of stress, temporarily worsens turf appearance and may create conditions favourable for invasion by pests.

## Soil Compaction

Compaction is defined as reduction in soil volume. Compressing forces from vehicular and foot traffic presses individual soil particles closer together, resulting in a denser soil mass. Compaction does not directly reduce turf-grass activity but rather affects its growth by influencing soil properties. The primary effect of soil compaction is reduction in pore volume and redistribution of pore sizes. These changes influence many other physical properties of soil, such as aeration, water retention, drainage and mechanical impedance to root growth. A reduced ratio of macropores to micropores may result in higher soil moisture content, but the water may be held so tightly that its availability to plants is limited.

Further, the reduction of pore diameters is often accompanied by loss of pore continuity, which reduces even further the diffusion of air and soil water movement and results in further growth reduction. Compaction also increases soil strength, especially when the soil is dry. The mechanical resistance of compacted soil significantly limits the growth of deep roots and therefore reduces turf-grass drought resistance.

Compaction of turf soils typically occurs close to the soil surface, and can usually be corrected by cultivation devices that penetrate the compacted soil relatively shallowly, not more than 10–15 cm. Several cultivation techniques can be used to alleviate surface compaction, but *core aerification*, sometimes simply called *coring*, is the most frequently used. Besides reducing compaction, coring also controls thatch accumulation, so one operation offers double benefit.

## Cultivation Methods

*Coring* is a cultivation method in which small holes are made in the soil, usually by removal of small cylinders or plugs of soil and turf. These plugs are extracted by means of devices with hollow tines, spoons or screws that pull soil to the surface. The most commonly used are vertical, self-powered hollow-tine aerifiers and tractor-pulled drum surface-coring units. The vertical-tine type is the most effective on high-maintenance areas such as golf greens. Tractor-pulled, spoon-type aerifiers are used most often on large areas such as golf course fairways or large athletic fields (Fig. 8.1). During operation, the cylindrical, hollow tines are first inserted into the soil, so that small cylinders of soil and turf are forced in through their open ends. When they are removed, they pull these cores out of the ground. The tines are then discharged; i.e. the cores are pushed out of the tines and left behind on the surface of the soil. These cores or soil plugs are usually 1–2 cm in diameter and 5–10 cm in length. They must either be removed or be broken up and worked back into the turf. The density of cores depends upon the type of machine but generally ranges from 100 to 200 per $m^2$. Core aerification can be quite disruptive to the surface, but within 1 week the turf area usually returns to its typical appearance.

Hollow-tine aerification is the most effective technique for relieving compaction. The removal of the core permits nearby soil to slide slowly down into the opening. Within 1 or 2 weeks, the core holes are filled with roots growing vigorously in a loosened soil (Fig. 8.2). The cores themselves are a valuable by-product; they are considered an excellent source of vegetative material for turf-grass propagation. When spread on a newly established seedbed, they require only moderate watering and, of all establishment techniques, assure the fastest and the most uniform

**Fig. 8.1.** Tractor-pulled, spoon-type aerifiers are used on large turf areas.

establishment of turf. Cores for propagation must be used when fresh, shortly after coring is performed.

Solid tines are sometimes used in place of hollow tines and can be mounted on the same machines. The solid tines penetrate the soil and open holes, but no soil plugs are removed. The solid tines are less disruptive to the turf surface, and recovery is faster, but they are less effective in loosening the soil and relieving compaction.

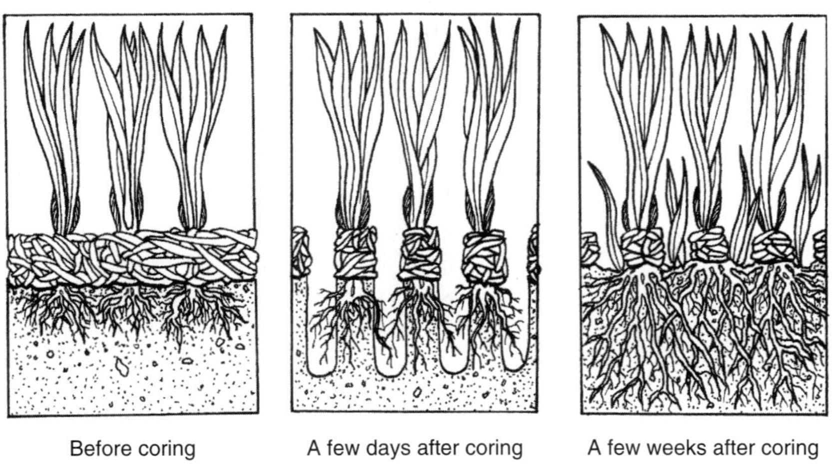


**Fig. 8.2.** Effect of coring on development of roots.

*Deep-tine* aerification helps to alleviate the problem of subsurface compaction (Fig. 8.3). Deep-tine machines are available with both hollow tines and solid tines, and are designed to penetrate to a depth of about 30 cm. The deep hollow tines are very effective at reducing compaction, but the process is more disruptive to the surface and recovery is slower.

*Slicing* is a cultivation method in which rotating flat tines or stationary blades slice vertically through the turf and soil (Fig. 8.4). Slicing is usually used on moist soil, penetrates to a depth of 10–15 cm, does not relieve soil compaction as efficiently as hollow-tine aerification, but causes much less surface disruption. Slicer blades are about 10 cm long, not power driven and relatively thin. Some units have wider and longer blades that can be adjusted to create a torque as they slice into the soil. On drier soil these blades produce a 'shattering' effect that is more effective in relieving soil compaction. On some units the blades are mechanically vibrated and able to shatter the soil to the depth of 20 cm.

*Spiking* is a cultivation method in which non-power-driven, solid, flat or pointed blades penetrate the turf and soil surface. The blades are similar to slicing blades but have smaller, knifelike tines and do not penetrate more than 3–5 cm into the turf. Spiking can be conducted often (even weekly) on moist soil.

**Fig. 8.3.** Deep tine penetrates soil to a depth of 30 cm, effectively reducing subsurface compaction.

**Fig. 8.4.** Rotating flat tines slice vertically through the turf and soil, reducing compaction and improving aeration.

*Subsoil* (or simply 'sub') *aerification* is a cultivation method in which bullet-shaped devices are forced through the deeper soil zones to break up compacted layers below the surface. Sub aerification is usually performed on moderately dry soil and requires a tractor generating a substantial amount of mechanical power.

*Forking* is a cultivation method in which a fork or similar solid-tine device is used to make holes in the turf. Forking is performed on high-maintenance turfs such as golf greens and some sport fields, especially those that have developed localized dry spots. The holes are usually slender and deep, up to 20 cm, and the surface is not disrupted.

*Water-injection aerification* is very effective, but also quite expensive to carry out. A special machine injects 'bullets' of water, under very high pressure, into the turf surface. The drops or streams of water relieve soil compaction without any disruption to the surface.

## Thatch Control

The second most important reason to cultivate turf-grasses is thatch control. Thatch is a layer of living and dead grass stems, roots and other organic matter that is found between the soil surface and the grass blades (Fig. 8.5). Its development indicates an imbalance between the amount of organic matter produced by the turf-grass plants and the rate of its decomposition. The major factors in thatch development are growth rate and growth habit of the turf-grass in question, as well as the maintenance of the turf area. In general, any turf-grass species that spreads by above-ground stolons or below-ground rhizomes is likely to produce thatch. For example, cynodon and zoysia (which both have stolons and rhizomes) are prolific thatch producers. These species respond actively to high fertilization, and under intensive levels of maintenance develop moderate to heavy thatch. In most turf situations, some amount of thatch is desirable. A moderate layer decreases damage to the area from excessive traffic. For example, highly trafficked turf areas on football, rugby or hockey fields may greatly benefit from thatch, which can also act as a valuable shock absorber where players frequently fall on the field surface. Excessive thatch, however, is disadvantageous.

**Fig. 8.5.** Thatch, a layer of living and dead grass stems, roots and other organic matter, develops between the soil surface and the grass blades.

It harbours potentially destructive diseases and turf insects, limits root penetration into the soil and, in some cases, causes hydrophobicity. Heavily thatched turfs may also affect games in which the bouncing response of the ball is critical to the performance of the players.

Any factor that influences the rate of production and/or decomposition of organic matter influences production of thatch. As the balance between accumulation and decomposition is swayed one way or the other, the thickness of the thatch layer increases or decreases. Major influences include grass species, fertilization (especially nitrogen), soil pH, mowing height, pesticide applications, soil type and in some turfgrasses means of clipping disposal.

Thatch is broken down primarily by the activity of soil-borne microbes and earthworms. Any factor that increases their activity accelerates thatch decomposition. Therefore, maintaining an appropriate pH, good soil aeration and adequate soil moisture reduce thatch accumulation. On the other hand, increased nitrogen fertilization usually speeds up turf-grass growth and results in faster thatch accumulation. Neutral soil pH and some natural organic fertilizers improve composition of the microbial community, therefore reduce thatch build up. Some pesticides increase thatch levels by adversely affecting the composition and activity of thatch-decomposing microbes. Mulching mowers, which finely chop grass blades, promote decomposition and therefore contribute less to thatch than do other types of mowers. In some species, like zoysia grass, the clippings contain high concentrations of silica, which protect the leaves from fast decomposition and therefore add to the thatch layer. Soil type may also affect thatch formation. Turf being grown on heavy soil usually develops thatch faster than the same turf grown on better-aerated sandy soil.

When thickness of the thatch layer exceeds 1.5 cm, it can become a problem and should be reduced. The thatch layer should be examined in several cores removed with a soil sampler or even a pocket knife. If its thickness significantly exceeds 1.5 cm, dethatching is recommended. One of the most common methods of thatch control is vertical mowing, sometimes called 'verticutting'. The blades of the vertical mower (also called a dethatcher) cut vertically into the turf canopy, severing lateral stems and removing thatch that develops on the surface of the soil (Fig. 8.6). Thatch is brought to the surface of the turf, from which it must be raked and removed. The tearing action of the blades can seriously disrupt the surface, so after vertical mowing, turf requires a period of recuperation, usually 2–3 weeks. Some turf areas never need dethatching, but others need it quite frequently. Golf greens in the tropics are dethatched every several weeks but lightly; recreational areas or home

**Fig. 8.6.** The blades of the vertical mower cut vertically into the turf canopy, severing lateral stems and removing thatch that develops on the surface.

lawns, especially when highly fertilized, may need extensive detaching every 1–3 years.

Like cores after aerification, dethatching debris contains many viable sprigs, which can serve as material for vegetative propagation. Dethatching should be performed at times of vigorous growth and should be avoided before anticipated periods of environmental stress, such as drought. After dethatching, fertilizer and water should be applied to promote rapid turf recovery.

## Top-dressing

In agriculture, *top-dressing* refers to application of a thin layer of manure or fertilizer, but in turf management, it refers to the distribution of a thin layer of soil over a turf-grass area. It is usually used to cover thatch or to smooth the playing surface. It is often performed in combination with other cultural practices such as core aerification and verticutting. High rates of top-dressing are often applied on newly established greens or athletic fields for the purpose of covering creeping stolons and smoothing the surface and are usually continued until the desired shoot density and smoothness are achieved (Fig. 8.7).

On already well-established and smooth turf, surface top-dressing should be used as infrequently as possible. If the thatch accumulation rate is very high, a better choice is to correct its fast build-up by modifying the fertilization programme, correcting soil pH, increasing soil aeration, and so on, rather than cover it routinely with top-dressing.

**Fig. 8.7.** Top-dressing, the application of a thin layer of soil over a turf-grass area, is usually used to cover thatch or to smooth the playing surface.

Even though top-dressing looks quite simple, it is one of the most difficult management practices performed on high-quality turfs. The greatest danger of this cultivation technique is soil layering, and selecting the right top-dressing material is the greatest challenge. When top-dressing is carried out incorrectly over a period of years, the entire turf surface may have to be removed and an expensive turf reconstruction conducted. When finer soil is laid over coarser soil, water movement can be seriously disrupted and turf growth severely impeded. In general, top-dressing soil should be of the same or slightly coarser texture than the underlying soil. So long as a detailed record is kept from the time of initial turf construction and is followed by the turf manager, top-dressing is sound and usually does not cause any long-term problems. If these records do not exist, or they are not followed, as often happens when turf managers change frequently, problems are virtually certain. Often, every new manager implements a new top-dressing programme. After several years, the result can be a sequence of contrasting layers of top-dressing material, each correlated with the time of manager's employment. If the textures of these multiple layers differ substantially, they may cause soil problems so severe that reconstruction of the entire area is unavoidable.

Top-dressing rates normally range from 0.2 to 0.5 $m^3$ of soil per 100 $m^2$ (a soil layer 2–5 mm thick). The applied soil should not cover the turf-grass leaves completely and should allow the turf-grass plants to carry out photosynthesis. Top-dressing is often applied by special machinery, but on small areas it can be distributed with a drop-type fertilizer spreader or even applied with a shovel and smoothed with a large broom.

## Rolling

*Rolling* is a method of smoothing the turf-grass surface. It should be used not to correct variations caused by improper levelling prior to seeding or sodding, but rather to press turf-grass plants back into the soil after they have been lifted upward, for example by vehicular traffic on an excessively wet surface. In the tropical climate, rolling also may be needed as a cultural practice for certain sports such as lawn bowling, tennis and cricket, for which a level, firm surface is required.

# 9 Golf-course Maintenance

A golf course can be defined as a large area landscaped for the playing of golf. On the golf course, turf-maintenance practices are largely determined by the requirements of the game rather than the requirements of the turf-grass plants. From an agronomic perspective, turf-grass mowing is generally excessive, fertilization is too high, cultivation practices are scheduled not when they are the most desired but so as to minimize disruption of play, the turf can be irrigated only when golfers are not on the course, pesticides are often applied on the basis of visual criteria rather than environmental demands, and so on. The turf manager must meet many challenges to provide the best possible turf surface for golf, and must, at the same time, satisfy the fairly high aesthetic expectations of the golfers. All these factors make proper golf course management a demanding and uneasy task.

Golf is an outdoor game during which each player moves a small ball from a starting point into a small hole in the turf, located from 120 to 550 m away (130–600 yards, as golf-course distances are traditionally given in English units), usually by striking it three to five successive times with specially designed clubs. The player repeats this process 18 times during one game (one round of golf). The many rules of the game will not be presented here, but some fundamentals are necessary to an understanding of the unique nature of golf course maintenance.

The term 'hole' is used to refer to the entire area from the starting point (the 'teeing ground', also called the 'tee') to the putting green (the area immediately around the actual hole in the turf that serves as the player's goal, also simply called the 'green'). The largest turf area between the tee and the green is called the fairway, and an area surrounding the tee, the green and the fairway between them is called the rough. Many holes also include 'hazards' such as 'sand traps' (also called 'bunkers') and bodies of water (lakes, ponds, rivers, etc.), collectively called 'water hazards'. The hole on the green is marked by a small flag on a pole so that golfers can see it from some distance. Most golf courses consist of 18 holes and cover an area of about 40 ha, although some courses have only nine holes, in which case each hole is played twice.

Every hole is classified by its par. Par, a number from three to five, is the number of strokes that a hypothetical expert golfer should require to play the ball into the hole. Par is determined principally by the distance from tee to green, on the assumption that golfer will need one, two or three strokes to cover this distance and then will need two rolled strokes ('putts') to reach the hole, once the ball is on the green. An 18-hole golf course typically consists of about four par-three holes, ten par-four holes and four par-five holes, for a total par of about 72.

## The Putting Green

Putting greens are the most important part of the golf course (Fig. 9.1). Even though they represent only 2–3% of the turf area, they require

**Fig. 9.1.** Despite their relatively small size, putting greens are the most important part of the golf course.

about 30% of the total turf maintenance. The greens are so important because, unlike the situation on tees, fairways and roughs, the golf ball remains on the surface at all times and must roll smoothly and uniformly and because half of the strokes in a game are expected to be putts.

Putting greens usually range in area from about 300 to about 700 $m^2$. Generally, greens that require a longer approach shot should be larger than those requiring a short approach shot. Therefore, long par-3 and par-4 holes should have the largest greens, whereas short par-3 holes and par-5 holes, on both of which the shot that lands on the green is expected to be played from a short distance, should have smaller greens. The putting green should absorb well the impact of a shot from the fairway and should provide a smooth, uniform roll of the ball at certain speed, called the speed of the green. The minimum required speed of a green is regulated by the system of golf rules, but in practical terms can be described as a measure of how far the ball should roll when started at a given velocity. Speed is an important component of green maintenance and is primarily controlled by mowing height, although other factors such as species and cultivars, mowing frequency and succulence of the grass also influence it.

### Sand-based putting greens

On many older golf courses around the world, the greens are constructed from natural soil. The majority of these courses are in the UK, Australia, South Africa and other countries in temperate climates. In the tropics, however, golf has been growing rapidly during the last several decades, so most of the golf courses are relatively new, and most of the greens are artificially constructed. At tropical golf resorts where many rounds of golf per day are played 12 months a year, the soil compaction on natural putting greens would be impossible to manage. A natural soil may become so severely compacted that alleviation of soil compaction would be only marginally effective and probably short lasting. Historical attempts to modify existing soil by mixing it with sand have proven ineffective. Researchers around the world have tried various methods and finally demonstrated that replacing soil with almost pure sand produces the best results. Compaction of these greens still poses a challenge, but a manageable one. Standard techniques of constructing greens have been developed and with some minor modifications have been adopted and implemented around the world. The most standardized green around the world is often called the *USGA green* because the United States Golf Association (USGA) first declared this design the industry standard.

A USGA green consists of a 30-cm-thick layer of sand mixed with 5% silt and clay and 5% organic material (peat moss), placed over a 10-cm-thick gravel layer (Fig. 9.2). This series of layers produces a perched water table, in which water drains only to the intermediate layer rather passing through it and draining away. Capillary forces hold the water in the fine pore spaces above the gravel against the force of gravity, where it can be used by turf-grass for a relatively long period of time. When heavy rainfall occurs, the excess water easily drains to the underlying gravel, where capillary forces are not sufficient to prevent it from draining away. When the rain stops, the weight of water in the sand layer again comes into balance with capillary forces, and drainage stops. This system of drainage is thoroughly described in Chapter 3. Note that only if the exact specifications for the USGA green are followed will its drainage function correctly.

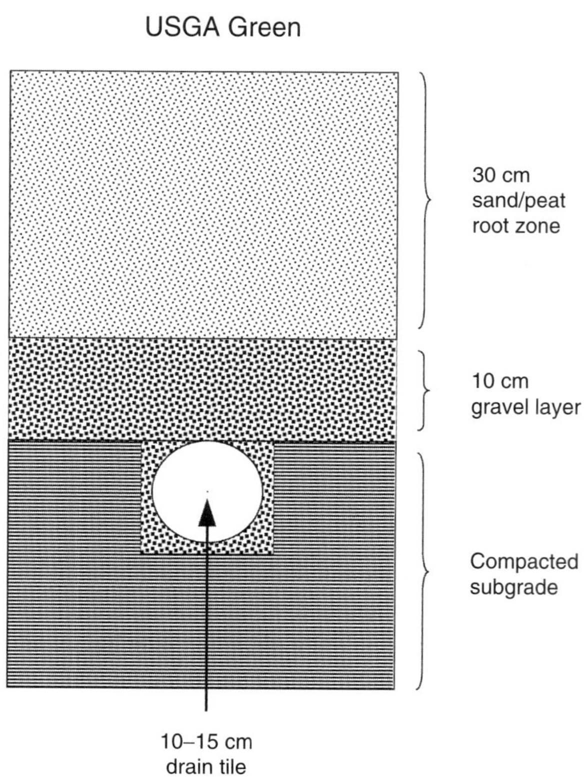


**Fig. 9.2.** Profile of a United States Golf Association golf green.

### Selection of turf-grass for putting greens

For tropical areas, turf-grasses desirable for putting greens include hybrid cynodon (*Cynodon dactylon* × *Cynodon transvaalensis*) and seashore paspalum (*Paspalum vaginatum*). Some golf courses in Singapore and Malaysia still use *Digitaria didactyla,* locally known as serangoon grass, which is closely related to a very troublesome weed, *Digitaria sanguinalis.* Numerous varieties of hybrid cynodon and paspalum are available around the world, and in certain countries sprigs may be imported or locally produced.

### Mowing

Putting greens are mowed very low, so even minimal mowing error or inconsistency in mowing height becomes noticeable. Before mowing begins, the green should be inspected for foreign objects such as stones, golf-shoe cleats, metal and other hard objects that may damage the mower-reel blade and/or bed knife. Ball marks should be repaired and smoothed, as they may cause scalping or tearing of the turf. During the inspection, the green should be checked for disease and insect activity, dry spots, wet spots, leaking irrigation valves, etc. The mower operator should take notes about the condition of each green and report findings to the turf manager when mowing is completed. For mowing putting greens, special high-quality mowers with 8–9 blades in the reel are required. Two basic types of mowers are used on greens: (i) a walking mower with a 50–56-cm mowing width; and (ii) a riding, three-gang (triplex) mower with a 1.5–1.6-m mowing width (Fig. 9.3). Walking mowers are more accurate but take 5–6 times longer. Riding triplex mowers are faster but usually cause more soil compaction and turf-grass wear, may be difficult to operate close to sloped bunkers, and sometimes cause damage to turf by leaking hydraulic fluid. Putting greens must be mown very low to yield the highest turf density and uniformity. The preferred cutting height varies from 3 to 7 mm, depending on turf species and anticipated use. The lowest cutting height, even lower than 3 mm, can be used for special professional tournaments, but ordinary golfers are usually satisfied with a height of 4–5 mm. A mowing height below that recommended for a given species and/or variety should not be maintained for longer than a week. Somewhat increased mowing height (over 5 mm) can be used when additional leaf surface is needed, e.g. to promote a faster recovery from stress or injury to the turf.

As a rule, putting greens in the tropics should be mowed daily in the early morning. Less frequent mowing results in loss of turf density, scalping, coarser leaf texture, unpredictable changes in ball-roll direction

**Fig. 9.3.** For mowing of putting greens and tees, walking mowers are preferred. They provide the most accurate cut and minimize the chances of accidental scalping.

and loss of speed. Mowing is sometimes omitted for 1 day after top-dressing, cultivation or fertilization; this practice does not negatively affect turf and may actually improve its vigour. The mowing pattern is typically changed at each mowing, so that wear is minimized and grain does not develop. Greens are mowed in four directions, changing clockwise. A line running through the green from the fairway marks the 12:00 o'clock and 6:00 o'clock positions. On a given day, all putting greens on the entire golf course might, for example, be mowed along lines parallel to the 12:00–6:00 o'clock line. The next day they would be mowed along lines parallel to the 3:00–9:00 o'clock line, the third day parallel to 1:30–7:30, then 10:30–4:30, and on the fifth day back to 12:00–6:00 line. Figure 9.4 shows another possible four-day rotation. When this method is used, mistakes are very infrequent.

During mowing, the operator must maintain straight lines of mowing and minimal overlap between different passes. The walking mower should be operated at a uniform speed that is low enough to avoid heel marks on the green from long steps. Turns should be smooth and gradual. Triplex riding mowers require a much wider turning area than walking mowers; otherwise serious injury to the turf may occur. Each mower operator must watch for gasoline, oil or hydraulic-fluid leaks. Even small amounts dripping on the turf surface may cause ugly dead spots. Clippings from putting greens are always collected in special baskets and should be disposed of at a composting site or dump area.

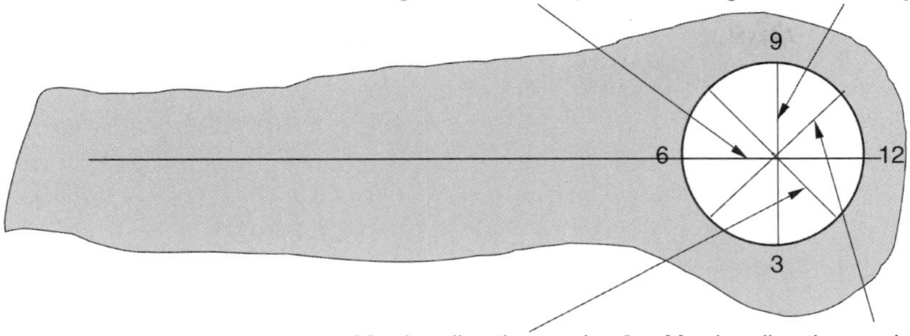


**Fig. 9.4.** So that grain does not develop, the mowing pattern on golf greens should be changed daily.

Despite great attention during mowing, turf scalping unfortunately happens from time to time (Fig. 9.5). On a putting green, it usually occurs because of excessive thatch build-up, excessively low cutting height or mistakes in mower height settings. If scalping is severe, mowing should be suspended in the scalped area for 1 or 2 days.

**Fig. 9.5.** The most common reasons for turf scalping are excessive thatch build-up, excessively low cutting height and mistakes in mower height settings.

## Fertilization

No uniform fertilization programme is suitable for all greens. Even different greens on the same golf course often have different fertilization requirements. A soil test should be conducted before a new fertilization programme is begun and all nutrient deficiencies corrected at this time. Afterwards, one of the following types of programme can be chosen. In the arid or semi-arid tropics, a complete fertilizer with a high nitrogen ratio (4-1-1) can be used throughout the programme. In the humid tropics, straight nitrogen sources can be used regularly and phosphorus and potassium supplied infrequently. A continuous supply of potassium in arid climates may somewhat increase turf-grass resistance to environmental stresses such as drought, heat and great changes between day and night temperatures (see potassium fertilization in Chapter 5, this volume). Because the level of environmental stresses in the humid tropics is usually lower, potassium fertilization can be more sporadic. Under either fertilization programme, a soil test should be done yearly that determines whether soil levels of essential elements are being maintained or whether adjustments are needed.

Because turf-grasses use nitrogen in large quantities, and because nitrogen readily leaches from the soil, nitrogen fertilizers should be applied routinely throughout the year. About 0.25–0.5 kg N/100 $m^2$ per month is required for cynodon putting greens. Paspalum requires less, about 0.1–0.2 kg N/100 $m^2$ per month. If a granular soluble nitrogen source is used, the total monthly nitrogen allotment can be divided in half and applied biweekly; if a soluble liquid source is used, weekly applications of one-quarter of the total monthly allotment should be used.

Fertilization with phosphorus, potassium and other essential elements was discussed in more detail in Chapter 5. Their management on putting greens is not different from that on other turf areas. No more than 100 g of iron sulphate, or ferrous ammonium sulphate per 100 $m^2$, should be dissolved in 25 l of water if iron is to be applied to the foliage, otherwise greens may experience burn or discoloration. If yearly soil tests detect any nutrient deficiencies, they should be corrected by application of the lacking nutrient.

## Irrigation

Irrigation of greens is challenging. Different green, or even different sections of the same green, may have different exposure, traffic intensity, shade conditions, slope, etc. All these factors influence water infiltration, percolation and retention, as well as evapotranspiration rates, and must be considered when irrigation is scheduled.

Irrigation systems installed on greens are usually underground, remote-controlled systems with rotary pop-up sprinkler heads. Generously budgeted golf courses use sprinkler systems in which each individual head can be controlled by a centrally located computer attached to a weather station. If such an irrigation system is not available, the turf manager must make water-scheduling decisions on the basis of readings from moisture-sensing devices, observation of the turf and professional experience.

Greens should be watered only to the depth of the upper sand layer. Irrigation should be scheduled to replace the water used by the turf when approximately one-half of that water has been used. Because water retention and water use will differ in different areas, the manager must closely and frequently observe the results and make necessary adjustments. Setting the irrigation programme 'once and forever' will not be sufficient.

### Cultivation and thatch control

Relieving compaction, improving water infiltration and percolation, controlling grain and thatch, and providing a smooth, uniform putting surface may require several cultivation operations, which were discussed in detail in Chapter 8. Here, only those modifications that apply specifically to cultivation of putting greens will be addressed.

A limited thatch accumulation of 5–7 mm increases wear tolerance, reduces soil compaction, and overall may be desirable for putting greens. Excessive thatch may cause mower scalping, reduced green speed, increased disease occurrence, development of hydrophobic areas and restricted water infiltration. Because thatch accumulates when its growth exceeds the rate of its decomposition, growth-stimulating practices should be minimized. Moderate nitrogen fertilization, supplemented with iron (which provides better colour), and use of slow-release nitrogen sources usually reduce rates of growth and thatch accumulation on greens. Maintaining good soil air – water relationships provides favourable conditions for microbial activity and also prevents thatch accumulation. Mechanical methods of thatch control are usually used routinely on sand-based greens and less frequently on natural greens. A thatch layer over 2.5 cm may require severe measures such as deep vertical mowing, frequent deep coring, or in extreme cases, stripping of the entire sod from the green and establishment of new turf.

Top-dressing is used on greens to control thatch, smooth the surface and control grain. On USDA greens, the top-dressing material is usually pure sand; on natural greens, soil or sand is used. As always the top-dressing material must be at least as coarse as the underlying soil.

Coring, slicing and spiking techniques for greens do not differ from those described in Chapter 8.

Rolling with water-filled rollers firms the root zone of newly established greens and smooths their surfaces. Recently, light-weight self-propelled riding rollers have become popular for smoothing the surfaces of putting greens and increasing green speed during tournaments. These rollers should be used only on sand-based greens. Soil greens may suffer excessive compaction even with light pressure.

## Cup changing

The rules of golf state that the hole on a putting green must be circular, not less than 4.6 m from the edge of the green, 10.8 cm in diameter and at least 10.8 cm deep. The slope should not change within 1 m of the hole. A metal or plastic cup is commonly placed inside the hole (Fig. 9.6). The cup should be sunk about 20-cm deep, and its upper rim should be at least 2.5 cm below the green surface. A slender rod (sometimes called a

**Fig. 9.6.** The cup should be sunk below the soil surface. A flag is usually placed inside the cup so that golfers can see the location of the hole from a distance.

'pin') topped by a small coloured flag is usually placed inside the cup so that golfers can see the location of the hole from a distance. In addition to these basic requirements, which usually satisfy recreational players, numerous additional stipulations must be followed during official golf tournaments but will not be addressed in this volume.

The location of the hole is usually changed daily to prevent localized excessive wear and shoe cleat marks on the green. The hole should be cut (with a cup-cutter) in an area as free as possible of ball marks, cleat marks or scuffed places. The cutting must be done with great care, and the sides of the hole must be perfectly perpendicular to the surface. The plug cut from the new hole must be placed in the old hole, and turf should be carefully trimmed around the edges. Light top-dressing can be placed around the perimeter of the plug, gently firmed with the hand and then watered.

The collar (sometimes called the 'fringe') is a band 1–1.5 m wide that surrounds the putting green and is mowed to a height intermediate between that of the fairway and that of the putting surface. It is considered an integral part of the green, and its maintenance, except mowing height, is the same as for the green.

## The Tee

The teeing ground is the starting place for the golf hole being played. It is usually rectangular, but can be of any shape. Two tee markers are placed on the surface to define a line behind which the golf ball must be placed for the initial stroke. A teeing ground can be of any size, but it should be large enough to allow for frequent relocation of the tee markers to prevent excessive localized wear. A problem on teeing grounds that greens do not share is *divots* (Fig. 9.7). A divot is a piece of turf that is torn up by the club as it strikes the ball; a divot opening is the resulting scar on the turf surface.

Because tees are subject to intense traffic, their soil should be very similar to that for putting greens. Grasses used on tees recuperate rapidly from wear, tolerate close mowing and provide a dense, tight turf, as well as tolerating soil compaction and wear. Cynodon, zoysia grass, paspalum and serangoon grass are well adapted for use on tees. In the tropics, *Zoysia japonica* for tees seems to be the best choice because of its excellent wear tolerance and dense, uniform surface. Seashore paspalum is also increasingly used.

Like putting greens, tees should be inspected before mowing for foreign objects. Tees are usually mowed lower than fairways but higher than putting greens. The tee surface should be relatively smooth, firm and uniform, and should provide a comfortable stance for the golfer.

**Fig. 9.7.** Numerous divot openings on the tee that have resulted from club strokes.

Mowing height for tees must be low enough that the turf does not interfere with contact between the ball and club when the ball is teed at the desired height. A low mowing height also provides a firm, stable stance for golfers. Heights of 7–10 mm are preferred. In the tropics, mowing frequency, as dictated by the rule that no more than one-third of the leaf area should be removed at each mowing (the 'one-third rule'), usually ranges from every day to every two days. Tees are commonly mowed with a triplex riding mower with six- to eight-bladed reels. Mower operation is similar to that previously described for mowing greens, but grain is of less concern on tees, so usually only two alternating mowing patterns are used.

Tees require somewhat less fertilization than greens, but sufficient nutrition must be maintained to provide rapid shoot growth to heal the extensive damage that results from divots.

Iron applications are quite important in tee fertilization. Monthly to biweekly applications of ferrous sulphate or ferrous ammonium sulphate provide excellent colour and increase wear tolerance of the turf grass. Iron application rates are similar to those used on the putting green. Other nutrients may be needed occasionally; the need should be determined from yearly soil-test results.

Tees should be irrigated somewhat less than are greens, and they should be kept somewhat drier to provide a firm stance for golfers and more tolerance to divot damage.

Tees need frequent cultivation to relieve compaction, control thatch and smooth the surface. Cultivation practices used on tees are the same as for greens. Frequencies of top-dressing and coring are usually lower than for greens, and vertical mowing should be deeper and less frequent than for putting greens.

Extensive divot repair is usually required. Golfers frequently repair divots themselves if a container of sand and a scoop are placed nearby. If the tee becomes extensively worn and damaged, areas may have to be taken out of play and resodded.

## The Fairway

The fairway is the turfed area between tee and the putting green. Fairway widths and lengths vary greatly depending on the total length of the golf course and its design. Fairways are frequently between 30 and 60 m wide and 100–500 m long, and they sometimes include features such as water hazards and bunkers (Fig. 9.8). The turf-grass on a fairway must be relatively dense, uniform, smooth and free from excessive thatch. The surface should be firm enough to provide a proper amount of roll after the ball lands on it and should not be grooved by maintenance equipment or golf carts.

The same turf-grasses adapted for use on tees are also used on fairways. In tropical areas fairways are usually planted with common cynodon, hybrid cynodon, *Z. japonica,* paspalum or serangoon grass. Hybrid cynodon has been the most common, but in recent years paspalum has received more attention, especially in locations irrigated with effluent or brackish water.

Fairways are mowed relatively low and quite frequently so as to provide the required density and firmness. Most turf-grasses on fairways are mowed to a height of 15–20 mm with reel-type self-propelled mowing units or gang mowers pulled behind tractors (Fig. 9.9). Mowing frequency is based on the one-third rule and in the tropics usually ranges from two to three times a week. Alternation of mowing pattern is encouraged but is not as critical as on greens or even tees. Mowers must not be driven at a speed that causes the mower to bounce. Bouncing, besides risking mower damage, produces a rippling effect in the mowing pattern that makes fairways look ugly. Turns should be made at reduced speed and in a wide arc so as to prevent tearing of the turf.

Unlike those on tees and greens, clippings on the fairway should be left in place. This practice reduces fertilizer requirements and usually

**Fig. 9.8.** Numerous bunkers are the most common type of hazard on the golf course.

does not interfere with aesthetic appearance. On fairways, the one-third mowing rule should be strictly followed so that clippings do not accumulate, shade the turf or interfere with play.

Fairway fertilization programmes depend on the turf-grass species or cultivars, the soil type, environmental conditions and cultural practices being used. A complete fertilizer with a high nitrogen ratio (4-1-1) can be used throughout the programme, or a complete fertilizer can be applied twice a year and nitrogen applied more frequently. Cynodon fairways generally require 15–40 kg of nitrogen per hectare per month, whereas zoysia grass, paspalum and serangoon grass require about 10–20 kg of nitrogen per hectare per month.

Nutrients can be applied as slow- or fast-release fertilizers or a combination of the two. The granular form is the most common, but liquid fertilizer formulations have been used successfully.

Fairways also benefit from applications of ferrous sulphate or ferrous ammonium sulphate, especially if the soil pH is above 6.5, conditions that sometimes create iron deficiencies. Deficiencies of other secondary nutrients and micronutrients are uncommon. Yearly soil tests should be conducted, and if a deficiency occurs, it should be corrected.

**Fig. 9.9.** Reel-type gang mowers pulled behind a tractor are one of the most popular types of mowers used on golf course fairways.

Irrigation of fairways is less critical than that of greens and tees. Because of differences in slope, shade, soil texture, etc., each fairway and each portion of the fairway should be irrigated separately. The sprinkler heads on the golf course should be grouped into zones of a few heads that water turf growing under similar conditions so that the run time can be most efficiently adjusted.

Irrigation should wet the rooting zone thoroughly, but should be as infrequent as possible. In arid climates, especially on shallow sandy soil, irrigation may be required daily or every 2 days. In the humid tropics, especially when soil is loamy, deep irrigation may be needed every 7–10 days. As explained in Chapter 7, the preferred time of irrigation is before dawn, but if the irrigation system has inadequate pumping capacity, watering throughout the night may be necessary.

The normal cultivation operations used on fairways include periodic coring and vertical mowing performed as described in Chapter 8. Top-dressing with soil or sand would be helpful, but on areas as large as fairways it is too expensive and is only occasionally applied.

Excessive numbers of divots frequently occur when turf is dense and closely mowed. Fairways should be continuously inspected by maintenance personnel and repaired as needed.

## The Rough

The rough is considered secondary turf on the golf course and receives the least maintenance. It is usually mowed at 50 mm or higher, irrigated moderately in arid climates and usually not at all in humid ones, and generally receives low levels of fertilization. Cynodon is the primary species used on rough, but zoysia grass, centipede grass, St Augustine grass, carpet grass and almost any other species could be used.

# Sports Turf

# 10

A sports turf area must be of a certain size and shape, appropriate to the sport that will be played on it. Its surface should be suitable to allow for fine competition, and its appearance should satisfy the aesthetic demands of the spectators. It is a challenging task to satisfy simultaneously the plants' needs for healthy growth, the players' wishes for a safe and uniform playing surface, and the spectators' desire for a visually attractive surface, especially when that surface must often be coated with colourful team logos, sports images, etc. In recent times, these challenges have continuously increased, making sports-turf management one of the most demanding lines of work in the professional turf industry. This chapter presents some of the sports played on turf, explain some principles of sports-field construction and point out ways in which their management differs from that of other turf areas.

## Rugby

Rugby is played in many tropical countries (Fig. 10.1). A fairly similar game called American football is played in very few locations, but the design and use, as well as the maintenance, of rugby and American football fields are very similar, so the information presented here applies to turf designed for both. In most other sports in which a ball is used, the turf surface must promote regular and predictable ball movement, but rugby (and American football) turf serves mostly as a platform for the movement of human bodies. The ball only infrequently touches the turf

**Fig. 10.1.** Rugby imposes heavy wear on turf-grass surfaces.

surface, so the turf's smoothness and uniformity are of secondary importance. Another unique aspect of these sports is the enormous force exerted by relatively heavy players on the turf under their feet. The players' shoes, equipped with long cleats, rip and tear the grass as players struggle for the ball or advantage. In addition, these large players often line up in the middle third of the field, away from the sidelines, so this portion of the turf sustains the most severe traffic stress. All these issues must be addressed in the processes of field construction, turf-grass selection and turf maintenance.

The dimensions of rugby and American football fields vary slightly, but most are approximately 100–120 m long and 50–60 m wide. Many fields are sloped toward one side (Fig. 10.2), but others are crowned longitudinally down the centre of the field, sloping toward both sidelines (Fig. 10.3). The major advantage of crowning is that surface-water drainage is directed away from the part of the field subject to the greatest traffic. On the other hand, drainage along a slope to one side is cheaper to construct and does not require installation of water-catching basins or other similar drainage structures. Usually, higher-budget fields are crowned, and lower-budget fields are sloped one way. In either case, the degree of slope should not exceed 1.5% and, in some cases, especially in the humid tropics, installation of a drainage system may be necessary.

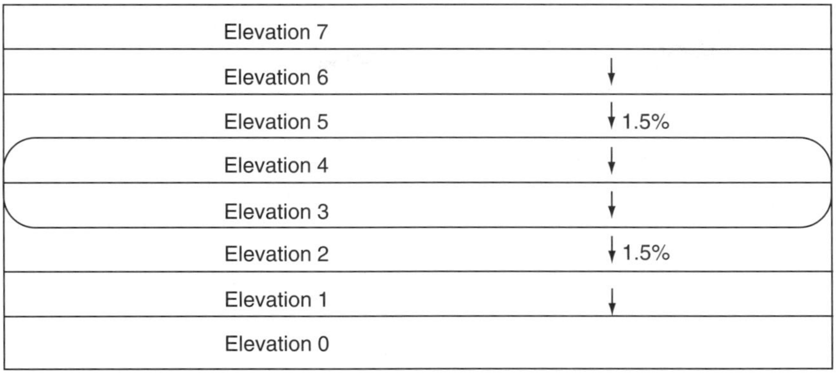


**Fig. 10.2.** A field sloped side-to-side provides water outlet towards one side only.

On rugby fields, drains are generally installed 50 cm apart and take one of two forms: the traditional pipe drain and the more novel strip drain. Pipe drains are usually laid 60–80 cm below the surface at the bottoms of ditches 25–30 cm wide and covered with gravel up to 15 cm below the surface. Topsoil is placed over the gravel, and turf-grass is established on the surface (Fig. 10.4). Strip drains are placed in shallower and narrow ditches, which are filled with sand to the surface (Fig. 10.4). Both types have advantages and disadvantages, so the choice should be made by an experienced drainage engineer on the basis of local conditions.

In tropical climates cynodon should be used on rugby fields. Cynodon is the only otherwise suitable species that tolerates extreme traffic, recuperates quickly from injury, and overall provides a long-lasting and visually appealing surface. Zoysia grass has better traffic tolerance, but its slow recuperation eliminates it from serious consideration.

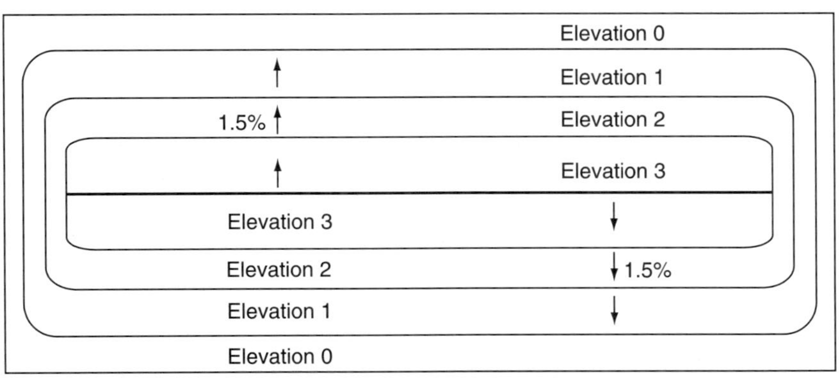


**Fig. 10.3.** A crowned field provides water outlet towards two sides.

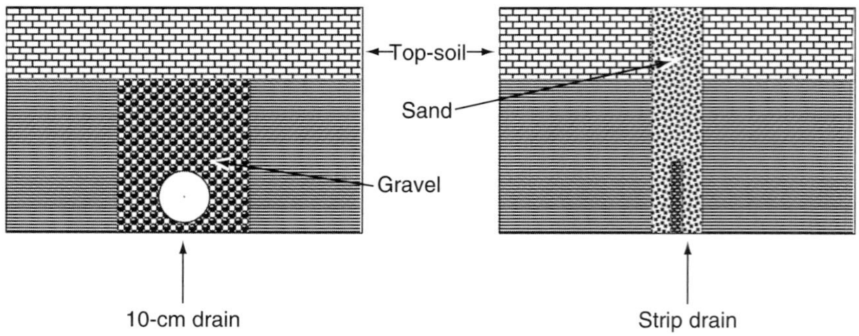


**Fig. 10.4.** A traditional pipe drain and a strip drain.

The cynodon on rugby fields is usually maintained under a very intense fertilization programme. Water-soluble ammonium nitrate, sometimes interchanged with urea, is the primary source of nitrogen. Nitrogen is applied frequently, at high rates, about 0.5 kg per 100 $m^2$ per month, and periodically supplemented with other essential nutrients. Mowing height is maintained at 20–30 mm, and when the one-third rule is followed (i.e. no more than one-third of leaf area is removed at each mowing), mowing frequency usually ranges from 2 to 3 times a week. Cynodon grown under such a high fertilization programme requires considerably more irrigation than do moderately maintained areas; its watering needs are typically higher by 50%. The size, durability and placement of sprinkler heads must take into account specific conditions of the rugby game. Sprinkler heads should be as small as possible, placed exactly at ground level and made from materials that will not be destroyed by athletes wearing shoes with cleats. The irrigation system must be designed so as to provide even distribution of water, but at the same time, sprinklers should not be placed in the areas where traffic is the heaviest. Like drainage, sports-field irrigation should be designed by experienced professionals.

During a time when games are not scheduled for at least several weeks, deep-tine aerification should be performed. Soil compaction, especially where players line up, is usually too great to be alleviated by ordinary coring with hollow tines, and sometimes deep-tine aerification must be performed. After aerification, top-dressing with sand is routinely recommended. Rugby fields may need periodic dethatching. Because some thatch on rugby fields is desired, dethatching once or twice a year is usually sufficient.

Post-emergence herbicides such as MSMA mixed with 2,4D should be used to control grassy and broad-leaved weeds. Herbicide treatments should be made during periods when games are not scheduled and also when no other environmental stresses are expected.

## Football and Hockey Fields

The planning, construction and maintenance of football (called soccer in the USA) and hockey fields are similar in many ways to those for rugby and American football fields, but they also differ substantially. Football and hockey balls continuously touch, roll and bounce on the turf surface (Fig. 10.5). Furthermore, players rapidly start and stop running, frequently change directions, deliberately slide to reach the ball with their feet, and jump to reach it with their heads. These major tactical elements of the game place a great deal of localized mechanical stress on the turf, but these stresses are less forceful and more evenly distributed across the field than in rugby. Football players are also much lighter than those in rugby or American football, and their shoes are fitted with shorter, less destructive cleats.

Football fields are about 90–120 m long and 45–90 m wide. Hockey fields are about 90 m long and 55 m wide. Most are sloped toward one side, but many are crowned longitudinally. In either case, football and hockey fields should not be sloped more than 1%, so surface water drainage is relatively slow and installation of underground drainage is often necessary. Effective drainage on football or hockey fields is especially important, because when turf is wet the ball rolls very slowly and rather than bouncing can be stopped dead in puddles of water. Especially

**Fig. 10.5.** In hockey, players frequently change directions, imposing wear stress on the turf.

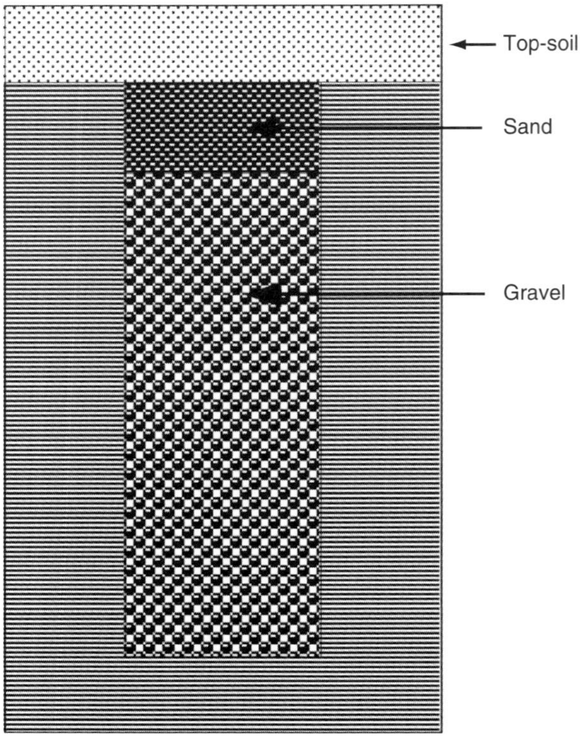


**Fig. 10.6.** A French drain.

in the humid tropics, where rains are heavy, the drainage system should absorb excessive surface water within minutes. Therefore, traditional pipe drains should be avoided; drains should be closer together than the commonly recommended 50 cm; or the so-called *French drains* (Fig. 10.6) should be used. French drains do not rely on pipes but rather on carefully constructed beds of gravel, which draw water away quite quickly and efficiently. Again all types have advantages and disadvantages, so an experienced drainage engineer should make recommendations based on specific local conditions.

Cynodon, the best choice for rugby fields, should also be used for football and hockey fields. It is the only species suitable for tropical climates that tolerates the traffic associated with these games, recuperates from injury quickly and provides a decent playing surface, although some faster-growing varieties of zoysia grass could be considered on school fields used predominantly by children.

On football and hockey fields, as on rugby fields, intense fertilization is necessary. Urea, the primary source of nitrogen, should be applied monthly at a rate of 0.5 kg of nitrogen per 100 $m^2$ or biweekly at 0.25 kg nitrogen per 100 $m^2$. Every few months fertilization should be supplemented with other essential nutrients.

Mowing height should be maintained around 20 mm, and under the one-third rule, frequency of mowing in most tropical locations ranges from 2 to 3 times a week. Clippings should be removed only immediately before a game.

Cynodon grown under high fertilization has higher irrigation needs than typical turfs. The size, durability and placement of sprinkler heads must take into account specific conditions of the game. Sprinkler heads should be as small as possible, placed exactly at ground level and made from a material that will not be destroyed by athletes wearing shoes with cleats. The irrigation system must be designed to provide even distribution of water, but sprinklers should not be placed in high-traffic areas. Like the drainage system, the irrigation system should be designed by a professional.

At times when games are not scheduled, soil compaction can be alleviated by spiking to the depth of 15 cm or by regular 10-cm-deep coring. Following core aerification, light top-dressing should be performed. Like rugby fields, football and hockey fields may need dethatching once or twice a year.

Post-emergence herbicides such as MSMA mixed with 2,4D should be used on cynodon infested with grassy or broad-leaved weeds. Herbicide treatments should be made during periods when games are not scheduled and also when no other environmental stresses are expected.

## Bowling-greens and Croquet Courts

Lawn bowling and grass croquet are played in many tropical locations, especially in the Commonwealth countries and others where British cultural influence is still significant (Fig. 10.7). The two sports use balls of similar size, both of which must roll smoothly across the turf surface. Both types of surface must be flat and relatively hard, so they are constructed and maintained in similar fashion. Compared to sports in which players run, rip turf with cleats or heavily crush turf-grass leaves, bowling-greens and croquet-grounds receive relatively little mechanical stress. Rather, the prevailing stresses are physiological in nature and are associated with very low mowing height and frequent water stress. A hard and dry turf surface mowed extremely low is needed for smooth and fast roll of the ball. Such turf is also frequently subjected to deliberate compaction with heavy rollers.

A lawn-bowling green is square, approximately 40 × 40 m. A six-wicket croquet-ground is 25 × 32 m, and a nine-wicket croquet-ground is 15 × 30 m. Both must be flat and well drained. Most are constructed from

**Fig. 10.7.** Croquet courts are mowed very low and are subjected to deliberate compaction and water stress.

fine sand similar to or even finer than that used to construct golf putting greens. Subsurface drainage is ordinarily required in all but arid climates. Because maximum firmness and the highest possible rolling speed are needed, the turf surface is kept as dry as possible and irrigated manually when necessary rather than automatically, on schedule.

Cynodon turf assures the best surface for both sports. Zoysia grass would be even better from a visual point of view, but presently available varieties cannot tolerate the desired 3-mm mowing height. Because the surface must be hard, thatch must be kept to a minimum by frequent light verticutting. Fertilization should also be kept low, sufficient only to maintain a reasonably pleasant appearance. Core aerification may be needed every 3 months, followed by light top-dressing. If the turf is under continuous use and cannot be aerified with hollow tines, less disrupting solid tines or water injection are preferred.

## Tennis Courts

Few tennis courts are now covered with turf – most are now paved – but in the tropical climate, especially when the weather is hot, naturally cool grassy courts are very appreciated and can still be found at tropical resorts.

Like those of bowling-greens and croquet-grounds, tennis court surfaces must be flat and relatively hard. Because slope should not exceed 0.5%, surface drainage is usually modest, and subsurface drainage is frequently required. Most tennis courts are constructed from fine sand and sodded with cynodon. Other species are seldom used because they grow too slowly or cannot tolerate the damage from intense foot traffic and ball bouncing. Mowing height is usually kept at 6–8 mm, and little thatch is allowed. The surface must be hard, but not nearly as hard as on bowling-greens or croquet-grounds, so water is only slightly restricted. The rate of nitrogen fertilization is relatively high, but much lower than that on football or rugby fields. Core aerification may be needed every 3 months and should be followed by top-dressing. Tennis facilities usually include a number of individual courts adjacent to each other. If courts are under extensive use, some of the individual courts should be placed off-limits for several days, in rotation, so that turf can recover and cultivation measures such as aerification, vertical mowing, top-dressing, etc. can be conducted. On extensively damaged areas, sodding rather than any other turf propagation method, should be used to repair bare spots.

## Playgrounds

Playgrounds are probably the most common sport/recreational facilities in the world (Fig. 10.8). They support a very broad range of play activities rather than a single sport. Stress imposed on the playground surface by traffic can be severe and frequently exceeds that imposed on football or rugby fields. Areas under swings and around slides and similar equipment are not only heavily compacted, but also frequently worn beyond the ability to maintain turf at all. Areas like these should be paved with some other material such as wood chips, rubber chips or sand rather than turf. Quick, effective drainage is of the greatest importance in maintaining playground facilities. Slightly sloped elevated areas with rapidly draining soils are the most suitable. The essential criteria for the turf species chosen are durability, density and safety. Turf playing surfaces must provide a soft and effective barrier separating the feet of playground users from the soil beneath. Children have a tendency to play just minutes after rainfall, so if drainage is insufficient and density inadequate, the children soil their clothes and shoes and track mud into houses, kindergartens or schools, distressing their parents or teachers. Turf-grass species selection is less critical, but turf must be dense and some thatch is desirable. Zoysia grass is an excellent choice for all types of

**Fig. 10.8.** Playgrounds are probably the most common sport/recreational facilities in the world.

playgrounds. St Augustine grass is preferable under heavy shade, and centipede grass also does well, except in areas of heavy traffic. Cynodon can be used in sunny locations, especially in arid and semi-arid climates. In the humid tropics, cynodon should be avoided because it thins significantly during the rainy season – coincidentally the muddiest period on playgrounds.

# 11 Pests

Pests are living organisms that interfere in some way with something we consider important. For example, to farmers, pests may include insects and diseases that damage their crops, weeds that compete with crops for nutrients and water, aquatic plants that clog irrigation ditches or rodents that feed on the grain. To homeowners, pests may include filthy cockroaches, annoying mosquitoes, wood-destroying termites, etc. Overall the presence of pests in the environment is widespread and natural. When the degree of their infestation is small, they are usually ignored, but when their number or activity exceeds a certain tolerance level, they are ordinarily restrained or controlled in some manner. Level of tolerance may be very different for different people or in different situations. Some people may consider one cockroach unacceptable, even though hundreds would not alarm some others. Similarly, one weed in the centre of a golf putting-green may not be acceptable, while hundreds of identical ones growing in the rough may not even be noticed.

In turf-grass situations, the presence of pests exceeds the level of tolerance when their number or activity disrupts quality or uniformity of the turf. In turf-grass pest management, pests are usually separated into major groups that include weeds, diseases and insects. Because a large variety of different organisms infests turf-grasses in various geographical locations or in specific management situations, this chapter discusses groups of pests and methods of their control rather than the characteristics of individual pest species. Species-specific information on particular pests and their control should be checked at local pest-control bureaux or places providing similar information. In some countries, selected pesticides mentioned or discussed here may be restricted or unavailable for various reasons. In such cases, the local pest-control bureau may suggest suitable alternatives.

## Pesticides

Pesticides are chemicals used to control pests. They are usually organic compounds that interfere in some way with one or more physiological processes in the pest organism. To be effective, they must affect the pest at some minimum concentration for a sufficient period of time. Effective pesticides are those that reduce pests' populations or their activity to a tolerable level, not necessarily that eliminate the pest completely. A few spots on the leaves or the presence of a few insects in the turf usually does not cause damage or reduce turf's quality below an acceptable level. Complete control – i.e. killing every single pest individual – even when it is possible, is usually unnecessary and would require higher doses of pesticides, which are both costly and harmful to the natural environment.

### Fate of pesticides

Pesticides are subject to many forces operating within the natural environment (Fig. 11.1). During application, a portion of the pesticide spray

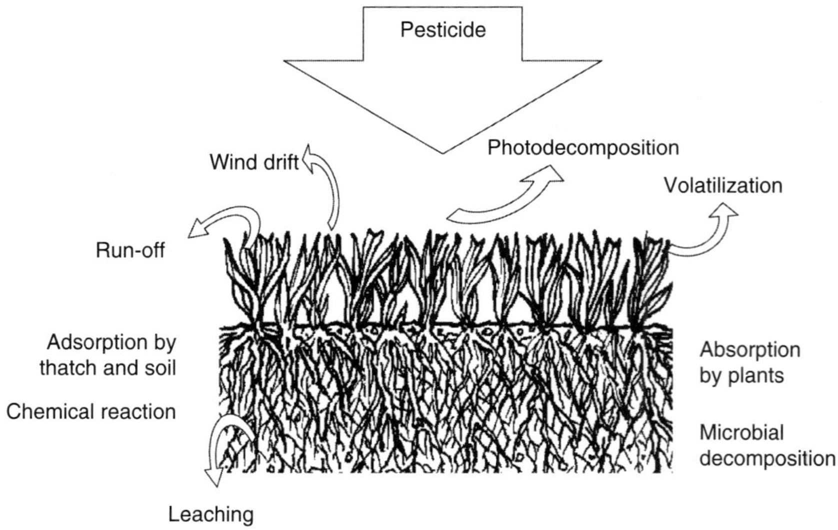


**Fig. 11.1.** Once released to the environment, pesticides undergo numerous transformations.

may be subject to wind drift and, in the case of some herbicides, may cause significant damage to nearby trees or other desirable plants. The potential for drift is reduced if pesticides are applied during periods of calm weather, preferably in the morning with correctly calibrated pesticide sprayers. Sprayers should operate under proper tank pressure and have nozzles of the correct size. If drift is likely, tank pressure can be reduced to make spray droplets larger and more resistant to drift.

Once applied, pesticides may be subject to run-off across the turf area. If rain is expected, pesticide application should be delayed because run-off of pesticides can cause harm to plants in the adjacent landscape, unnecessary contamination of the natural environment and loss of pesticide.

Some pesticides can be adsorbed or bound to soil particles or thatch. Application rates of these pesticides may have to be increased when they are applied to soils containing large proportions of clay or organic matter.

Some pesticides are volatile. Especially at high temperatures they can easily change from a liquid to a gaseous form and be lost to the atmosphere. The vapours can then travel downwind, causing damage to sensitive plants located nearby. Watering them into the turf after application can reduce losses from volatilization of soil-applied pesticides.

Pesticides that are not readily adsorbed to soil particles may leach downward, especially in sandy soils. Leaching may carry pesticides down to the water table, where they can contaminate wells, ponds, streams, etc.

Once released to the environment, pesticides often undergo chemical transformations. For example, some foliage-applied pesticides are subject to photodecomposition (breakdown in sunlight). Because little can be done to prevent the process or slow it down, photodecomposition is usually taken into account when application rates are developed.

Virtually all types of pesticides applied to the soil undergo microbial degradation. Organic molecules of pesticide are used by microbes as a food source and degraded to relatively simple inorganic compounds such as carbon dioxide and other compounds naturally occurring in biological systems. Microbial degradation is usually much faster in the tropics than in cooler climates. As a result, the duration of residual pesticide activity is shorter, disappearance from the natural environment is quicker, and the potential for leaching to the groundwater is less.

Pesticides can be absorbed by the plants to which they are applied, and in the case of systemic pesticides, they can then be translocated to other parts of the plant. Translocation can permit control of insects or diseases otherwise protected from contact with the pesticide, e.g. those inside plant tissues. Stressed plants, however, may not be able to absorb enough pesticide to control the pest, so only actively growing plants should be sprayed.

## Pesticide labels

Just like medicines in pharmacies, all pesticides in the world must be sold with a document called a *pesticide label* attached (Fig. 11.2). This label provides basic information about the pesticide in the container, tells how to use it correctly, and warns of any environmental or health hazards associated with it. The laws of virtually every country regulate the type of information that must be included on pesticide labels. Some countries impose specific regulations, but most pesticide labels around the world would include:

- trade or brand name of the pesticide;
- manufacturer's name;
- pesticide's common and chemical names;
- type of pesticide (herbicide, insecticide or fungicide);
- type of formulation;
- a signal word indicating the pesticide's toxicity class;
- directions for use.

Most countries would also require information about the appropriate first aid after accidental exposure to the pesticide:

- storage and disposal information;
- the address of the manufacturer;
- product registration number.

The brand name of pesticide is usually printed in larger letters on the front of the pesticide label to identify the product on the market. Frequently, the same product, intended for the same use and manufactured by the same company, is sold under different brand names in different countries. Similarly, generic products containing the same chemicals are often sold under various names, so that users can compare these products, the chemical and/or common name of every active ingredient and their percentages by weight must appear on the label. In some countries, the common name, e.g. 'dithiopyr', is sufficient; in others the chemical name, e.g. S,S-dimethyl 2-(difluoromethyl)-4-(2-methylpropyl)-6-(trifluoromethyl)-3,5-pyridinedicarbothioate, must follow the common name. Names of active ingredient(s) are usually printed just above or below the information indicating the type of pesticide. The formulation of a pesticide identifies the way in which the active ingredient is mixed with inert ingredients to make it suitable for use. Common formulations include:

- emulsifiable concentrate (an oil-based liquid to be mixed with water before spraying);
- liquid (liquid to be mixed with water before spraying);

DowElanco

**Dursban***

Turf Insecticide

**Specialty Insecticide**

**To be applied only by or under the direct supervision of commercial applicators responsible for insect control programs. Sale to or use by persons owning or occupying a dwelling is strictly prohibited.**

Active Ingredient:
chlorpyrifos: O,O-diethyl O-(3,5,6-trichloro-2-pyridinyl) phosphorothioate ..................44.9%
Inert Ingredients: ......................................55.1%
Contains 4 pounds of chlorpyrifos per gallon.

EPA Reg. No. 62719-35 EPA Est. 464-MI-1

**Precautionary Statements**

**Hazards to Humans and Domestic Animals**

**Keep Out of Reach of Children**

**WARNING AVISO:**

**Precaucion al usuario:** Si usted no lee inglés, no use este producto hasta que la etiqueta le haya sido explicada ampliamente.

**May Be Fatal If Swallowed • Absorption Through Skin May Be Fatal • Causes Substantial But Temporary Eye Injury • Causes Skin Irritation**

**Do not get in eyes, on skin or clothing. Wear eye protection. Avoid breathing vapors and spray mist. Handle concentrate in a ventilated area. When handling concentrate, wear protective clothing such as long-sleeved shirt, long-legged pants, hat, rubber gloves and footwear resistant to aromatic solvents, i.e. neoprene or nitrile butadiene rubber. Wash thoroughly with soap and water after handling and before eating or smoking. Remove contaminated clothing and wash before reuse. Keep away from food, feedstuffs and water supplies.**

**First Aid**

**If swallowed:** Call a physician or Poison Control Center immediately. Do not induce vomiting. Contains an aromatic petroleum solvent. Do not give anything by mouth to an unconscious person.

**If on skin:** Immediately wash with plenty of soap and water. Get medical attention.

**If in eyes:** Flush with plenty of water for 15 minutes. Get medical attention.

**If inhaled:** Remove to fresh air if symptoms of cholinesterase inhibition appear and get medical attention immediately.

**Note to physician:** Chlorpyrifos is a cholinesterase inhibitor. Treat symptomatically. If exposed, plasma and red blood cell cholinesterase tests may indicate significance of exposure (baseline data are useful). Atropine, only by injection, is the preferable antidote. Oximes, such as 2-PAM/protopam, may be therapeutic if used early; however, use only in conjunction with atropine. In case of severe acute poisoning, use antidote immediately after establishing an open airway and respiration.

**Environmental Hazards**

This pesticide is toxic to birds and wildlife, and extremely toxic to fish and aquatic organisms. Do not apply directly to water. Drift and runoff from treated areas may be hazardous to aquatic organisms in adjacent aquatic sites. Cover or incorporate spills. Do not contaminate water by cleaning of equipment or disposal of waste. This product is highly toxic to bees exposed to direct treatment or residues on blooming crops or weeds. Do not apply this product or allow it to drift to blooming crops or weeds if bees are visiting the treatment area.

**Physical or Chemical Hazards**

**Combustible - Do not use or store near heat or open flame.**

**Notice:** Read the entire label. Use only according to label directions. **Before buying or using this product, read "Warranty Disclaimer" and "Limitation of Remedies" elsewhere on this label.**

In case of an emergency endangering health or the environment involving this product, call collect 517-636-4400.

Agricultural Chemical: Do not ship or store with food, feeds, drugs or clothing.

**Directions for Use**

It is a violation of Federal law to use this product in a manner inconsistent with its labeling.
Read all Directions for Use carefully before using.

**Fig. 11.2.** The pesticide label provides basic information about the pesticide in the container.

- soluble powder (powder to be dissolved in water before spraying);
- granules (to be applied dry);
- solution (ready-to-use liquid);
- aerosol (ready-to-use spray).

The label must also include a signal word and sometimes symbol that states how toxic the pesticide actually is. Signal words vary slightly, but in any language they must indicate the relative level of health risk the product presents. Typical words are 'highly toxic' (in many countries indicated also by the skull-and-cross-bones symbol), 'danger', 'poison', 'warning', 'caution', etc. Pesticide toxicity is usually expressed as *lethal dose value* or $LD_{50}$, which is the amount of the material in milligrams per kilogram of body weight required to kill 50% of a test population of laboratory animals. Pesticides with low $LD_{50}$ are the most toxic. For example, the soil fumigant methyl bromide has an $LD_{50}$ value around 2, whereas that of the herbicide benefin is above 5000. These figures means that a human weighing 50 kg who ingested 100 mg of methyl bromide would have only a 50% chance of surviving (2 mg/kg × 50 kg = 100 mg). The same human would have to ingest 0.25 kg of benefin to suffer the same risk (5000 mg/kg × 50 kg = 250,000 mg or 0.25 kg). $LD_{50}$ values are usually not printed on the label but are reflected in the signal words that appear there. The term 'highly toxic' usually corresponds to an $LD_{50}$ between 0 and 50, 'moderately toxic' to one from 50 to 500, 'modestly toxic' to one from 500 to 5000, and non-toxic to one above 5000. If the product is harmful to swallow or inhale or can damage eyes or skin, the label often includes emergency first-aid measures and explains exposure conditions requiring medical attention. In many countries labels also include instructions about avoiding harm to the environment, like such statements as 'Toxic to bees' or 'Do not apply where run-off is likely to occur'.

Most pesticide labels include directions for use. This section provides information on how to use the product properly within legal requirements to get the best results. The directions for use state what pests the given product is registered to control, give the form in which the product should be applied, list and describe required application equipment, provide mixing instructions, recommend a frequency of application, and so on. The law sometimes requires that the producer's name and address be printed on the label, along with an establishment number identifying the factory that made the chemical.

## Applying pesticides

Most pesticides used in turf-grass management are applied in the form of liquid sprays or solid dry granules. Because methods of pesticide application can affect human health as well as the environment, in many countries the law regulates pesticide application procedures. Some pesticides that are especially harmful to humans and the natural environment can be applied only by trained and licensed pesticide applicators.

The pesticide label specifies the proper methods of application for the product.

For effective control of turf-grass pests, pesticides should be applied uniformly and at the proper rates. Liquid formulations are applied in water by sprayers, which must be calibrated to deliver the appropriate volume per unit area. Granular products are usually applied with the same spreaders used for seeding or application of fertilizers (see Chapter 4, this volume). The choice of application equipment may depend on pesticide formulation and the type of area to be treated. Large, power-driven equipment may be chosen for large areas such as golf courses; small hand-operated equipment may be chosen for a home lawn.

*Sprayers*

Most pesticides on turf areas are applied as liquids with low-pressure sprayers, which normally deliver spray volumes ranging from 250 to over 1000 l/ha at a system pressure of 1–4 atm. Many different types and sizes of turf sprayers are available, varying from small units to machines with booms 10 m long or more (Fig. 11.3). Regardless of size, the majority of sprayer components are similar and usually include:

- the sprayer tank, which holds the liquid pesticide;
- a pump, which moves the pesticide from the tank to the nozzle;
- agitators, which mix the components of the spray mixture uniformly and, for some formulations, such as powders, keep the material in suspension;
- strainers, which filter the spray mixture, protect the working parts of the spray system from particulates, and prevent clogging of nozzle tips;

**Fig. 11.3.** Boom-type sprayers assure the most uniform distribution of pesticide over the turf area.

- hoses and a boom, which connect the tank and pump to the nozzles;
- pressure gauges and regulators, which monitor the spray system;
- nozzles, which break the liquid into spray droplets.

Of all these components, the nozzles are the most important, and the purpose of all other components is mainly to ensure proper operation of the nozzles. A nozzle has four major parts: (i) the nozzle body; (ii) strainer (screen); (iii) cap; and (iv) tip (Fig. 11.4). Successful spraying depends on correct selection of nozzle type and size. Nozzles come in three basic types, which deliver liquid in the shape of a solid stream, a flat fan or a cone. In addition each nozzle can apply liquid in a narrow, regular or wide-angle patterns. In turf-grass situations, the principal type of nozzle is the 80° flat fan. Wide-angle (110°) nozzle tips are especially useful in windy locations subject to spray drift. Nozzles should be kept clean, and the tips should be inspected before each use.

Hand or back-pack sprayers can be used to treat small or hard-to-reach areas (Fig. 11.5). Hand sprayers use compressed air to force the spray liquid through a single or multiple nozzles. The capacity of hand sprayers generally ranges from 5 to 20 l.

*Granule spreaders*

The most common devices used to apply granular pesticide formulations are drop (gravity) spreaders and rotary spreaders, already discussed in Chapters 4 and 5. Both types normally consist of three main components:

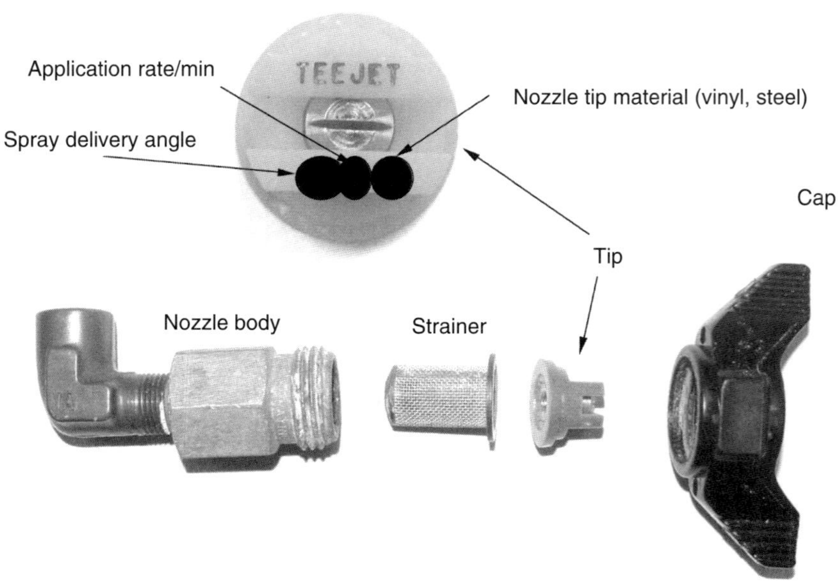


**Fig. 11.4.** The nozzle is the most essential component of the sprayer.

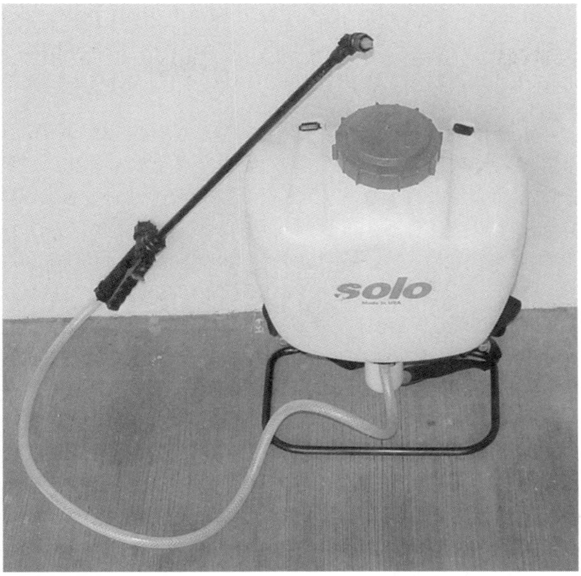


**Fig. 11.5.** Hand sprayers are used on small or hard-to-reach areas and usually have only one nozzle.

(i) a hopper to hold the pesticide; (ii) a mechanical agitator at the base of the hopper to provide a uniform and continuous feed; and (iii) some type of metering device, usually a slit gate, to regulate the flow of granules.

Drop spreaders, also called gravity spreaders, apply coarse, uniformly sized dry particles to soil and sometimes foliage. An adjustable sliding gate opens the outlet holes, and the granules flow out by gravity feed. A revolving agitator is activated when the spreader is in motion to ensure uniform dispensing. Drop spreaders are light and relatively simple and are generally more precise and deliver a better pattern than rotary spreaders. Because the granules drop straight down, chemical drift is minimal, control is good and the chance of applying pesticide to non-target areas is small. Because the edge of a drop-spreader pattern is sharp, any steering error will cause skips or overlaps. Application in crossing directions usually helps to achieve better coverage, but completely uniform application is difficult to achieve with a drop spreader.

Rotary spreaders, both power- and hand-driven, distribute granules to the front and sides of the spreader by means of a spinning disc or fan. Most rotary spreaders produce a belt width of 2–3 m, so they cover an area faster than drop spreaders. The rotary spreaders are usually easier to push than drop spreaders but are less precise in uniformity and distribution. In addition they are more sensitive to ground speed, physical characteristics of the granules and environmental factors such as wind.

A change in ground speed changes the distribution pattern; granules are thrown farther at high speeds and less far at lower speeds. Heavy granules do not travel as far as light granules, and drift may be a problem with fine particles when wind is present.

## Weeds

Weeds are plants that grow where they are not wanted. In turf-grass situations, a plant is usually considered a weed when its size, shape, growing habit or colour disrupt the uniformity of the turf area. Even normally desirable plants such as cynodon, which form attractive and highly desirable turf when grown alone, may be considered weeds if they grow in a turf of some dissimilar species, like St Augustine grass. The plant species considered weeds in turf are not generally the same ones considered weeds in agronomic crops. For example, broad-leaf weeds (dicotyledons) are a serious problem in agronomic crops but in turf pose relatively minor problems. Their growing points (apical meristems) are at the top of the stem and can be removed at each mowing. Therefore, most broad-leaf species can easily be mowed out, and only a few, those that have adapted to continuous defoliation, present a challenge if chemical control is not used (Fig. 11.6). On the other hand, 'grassy' species (grasses and grasslike plants, monocotyledons) can cause profound problems in

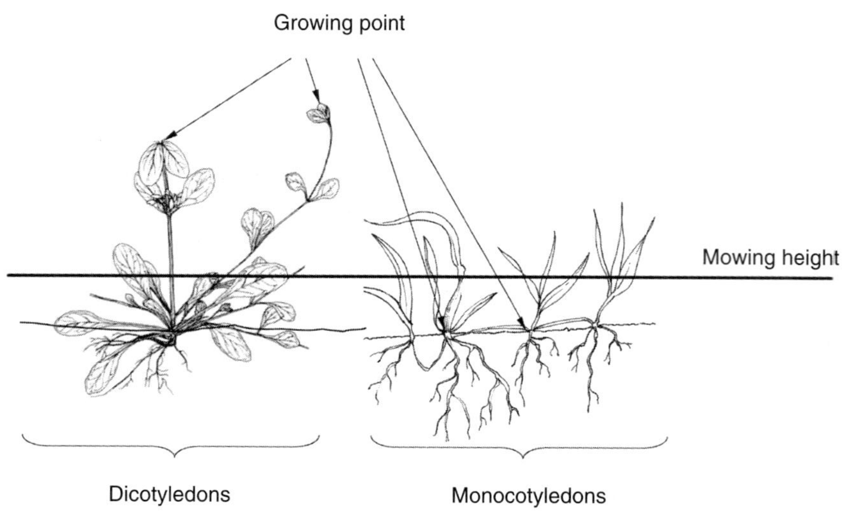


**Fig. 11.6.** Mowing cuts off stems and growing points of dicotyledons, and continually regenerating them depletes food reserves, whereas monocotyledons lose only leaf area and are little affected.

turf. Because these plants, like desirable turf-grasses, have growing points close to ground level, mowing does not eliminate them effectively. In the tropical climate, the most challenging are perennial and intrusive grasslike species that tolerate low mowing and turf-grass species encroaching on areas of other turf-grass species. Quite often, cultural control methods are ineffective and chemical control is required. In some instances even chemical weed control is insufficient, and in extreme situations the entire turf area must be renovated or the invaded turf-grass species must be replaced with another one.

A few terms used to describe weeds require clarification because, although they may not apply to the majority of weeds in tropical climates, readers may come across them while reading herbicide labels, turf publications, weed-science books, brochures or flyers. Weed scientists classify weeds into three major categories: annual, biennial and perennial. These terms are most relevant in temperate climates, where plant life cycles are more closely tied to the annual cycle of warm and cold seasons. Weeds classified as summer annuals germinate in spring when soil temperature reaches a certain level and usually die after the first autumn frost. Winter annuals emerge in the late summer and die the next year. Biennials live for two years only, accumulating resources in the first year and flowering in the second, and perennials persist for many years. Weed scientists at research centres and numerous chemical companies have developed weed-control programmes for their largest markets, most of which are located in temperate climates. Recommendations are developed for specific groups of weeds persisting in these climates and usually tell turf managers what cultural practices to employ, what herbicides to use, and how and when to apply them to assure the best weed control. In the tropics, however, many of these recommendations are not applicable. Annual weeds do not exist; species considered annual in the temperate zone persist for years, developing tougher stems. In dense turf, they spread mainly vegetatively rather than by seeds, and when established persist indefinitely. Weed-control data from temperate climates and herbicide label instructions should serve as general guidelines, but substantially different results can be expected. Information on weed control in the tropics should be drawn from publications prepared specifically for tropical locations.

### Non-chemical weed control

Healthy and vigorous turf is often free of weeds. Because weeds encroach mainly on weak or damaged turf, proper management techniques can stop or reverse weed infestation. Prevention is as important as cultural, mechanical, biological and chemical means of weed control.

The aim of prevention is to minimize weed introduction within the turf area. Weeds are prolific seed producers and can easily be spread by wind, water and animals or with contaminated planting material. Use of weed-free turf seeds, stolons, sprigs, plugs or sod is the most basic defence against weeds. As mentioned in Chapter 3, most national or local governments set limits on the contamination of seeding material. The limits vary from location to location, but most are below 1%, and all commercially distributed seeds must bear a seed label that provides the actual percentage of weed seeds in the package. In many countries, similar regulations with respect to vegetative materials are also established and enforced. Sod produced by sod farms is monitored by appropriate national or international sod-producers associations and in most cases should not contain more than 1 weed per 1 $m^2$, but in some cases no weeds are allowed. For example, cynodon in St Augustine grass sod is completely forbidden because after establishment no cultural, physical or chemical method will eliminate it.

Another main form of weed prevention is sanitation practices. Mechanical devices, such as mowers or cultivators, easily transport weeds from place to place, especially when turf is wet. As a general rule, less weedy turf should be mowed before weedier areas, and the equipment should be rinsed before being transported from weed-infested to weed-free areas. Frequently, weeds are introduced into the turf through the addition of contaminated topsoil. When weed-contaminated topsoil is used for top-dressing, adequate chemical control should be conducted before weed seeds germinate. Another contributor to the spread of weeds is irrigation systems drawing water from open ditches or ponds. Areas around ponds should be mowed and kept as free of weeds as possible.

Cultural practices that encourage robust and healthy turf are a very important but often neglected way of keeping turf free of weeds. Because many weeds need intense light for germination, high turf density may reduce their germination by preventing sufficient light from reaching the soil surface. All cultural practices that promote vigorous growth of turfgrasses, such as proper soil moisture, adequate fertilization, favourable pH and good aeration, help to make turf resistant to invasion by weeds. By the same token, any major stresses such as excessively low mowing height, insects, diseases or excessive shade, reduce turf density and allow weed encroachment.

Other significant means of weed control are tillage before turf establishment, hand pulling, hoeing and mowing. These methods are often categorized as mechanical and, especially when supplemented with cultural practices, provide quite satisfactory results.

Tillage, or cultivation, is usually conducted before turf-grass establishment. Weeds are removed or buried by the process, or are uprooted and die of desiccation. Tillage produces the greatest need desiccation if performed when the soil surface is dry and repeated several times.

Manual weed control, i.e. hand pulling or hoeing, has not been widely practised except on small residential lawns or other similar areas. Besides being very labour intensive, hand pulling and hoeing are inadequate when weeds are well established. Many broad-leaf and grassy weeds produce underground reproductive parts such as bulbs, tubers or rhizomes that remain in the soil and can quickly regenerate. On the other hand, proper mowing practices are a very effective tool of weed control. When weeds are mowed frequently, their reproductive organs are cut off before maturation, and their existing food reserves are depleted in the continuous regeneration of cut-off stems, so the plants eventually die. Most tall broad-leaf weeds can be completely eliminated by mowing, and flat-growing (prostrate) broad-leaf species and those that form rosettes can be strongly suppressed. Many prostrate dicotyledons have primary growing points located close to the soil surface. In these plants, mowing removes the oldest portion of the leaves but does not injure the growing point, so new growth can resume. Weeds tolerant of mowing include selected species of the families *Portulacaceae*, *Euphorbiaceae*, *Rubiaceae* and *Polygonaceae*. Many of these must be controlled chemically, otherwise their eradication from turf is virtually impossible.

## Chemical weed control

Cultural control can be effective in many situations, but under some circumstances chemical control becomes necessary. Because the major goal is the best control of unwanted weeds with the least harm to the natural environment, chemicals and the processes used to apply them should be chosen carefully. Information regarding proper selection of herbicides is usually provided by local pest-control bureaux, agricultural universities, research stations, etc. Basic information regarding proper use of chemicals is printed on the pesticide label, discussed above. The information on that label represents regulatory law. Certain modifications are allowed, but the turf manager must not cross boundaries of law.

### *Types of herbicides*

Over many years of weed science research, hundreds of herbicides have been discovered. They have been classified and grouped on the basis of their properties, chemical structure, mechanisms of their action, solubility and other factors. Only selected classes and few chemicals having wide application in turf-grass management will be mentioned here.

Herbicides are chemicals that kill or alter the normal growth of weeds. On the basis of their selectivity, herbicides are usually divided into two main groups: *selective* and *non-selective*. Selective herbicides control target weeds without harming desirable turf-grass species. Non-selective herbicides kill all vegetation, including turf-grasses. Selectivity of a herbicide is influenced by the way it is absorbed or translocated within the plant as well as by morphological or physiological differences between turf-grasses and weeds. The majority of herbicides for use on turf are selective. Non-selective herbicides are generally used instead in the renovation or establishment of new turf areas or as spot treatments of isolated weed patches.

On the basis of the timing of their application, herbicides can be classified as *pre-emergence* or *post-emergence*. Pre-emergence herbicides are applied before germination of weed seeds, and post-emergence herbicides are used to control weeds that have already emerged from the soil. In temperate climates, pre-emergence herbicides are important and very effective in controlling annual grasses and certain annual broad-leaf weeds. They act by forming a 'chemical barrier' in the soil that prevents seedlings from emerging and developing normally. The timing of application is the most critical component of their effective use, but they are a good choice when the majority of weed seeds germinate within a relatively short period of time. Nearly all seeds in temperate climates germinate in spring, when soil temperature reaches 13°C to 16°C. Pre-emergence herbicides remain effective in the soil for 2–3 months in the temperate zone and can therefore provide excellent control if applied just once a year. This pattern is not effective in tropical turf-grass management, because in the tropics, the same weed species act as perennials, germinate year round and persist in turf for many years. Under the tropical conditions of high moisture, high temperature and overall high microbial activity, herbicides stay active only for 4–6 weeks rather than 2–3 months and must therefore be applied many times per year. Because multiple applications are both expensive and environmentally unacceptable, use of pre-emergence herbicides in the tropics is quite marginal, and the majority of weeds are controlled after emergence.

Post-emergence herbicides are applied directly to growing weeds. In contrast to pre-emergence herbicides, this group does not provide any residual control of weeds. Because seeds germinate continuously, complete chemical weed control usually requires several applications per year. In many cases, turf discoloration or injury results from application of post-emergence herbicides that are not adequately selective. Some turf-grass species, especially when well established and mature, are quite resilient, but others are very susceptible to many post-emergence herbicides. Turf-grass plants that are still in juvenile stages are much more susceptible than mature plants. In general, regardless of turf-grass species, application of post-emergence herbicides on newly established sites should be delayed.

Post-emergence herbicides can be either *contact* or *systemic*. Contact herbicides affect only those parts of the plant that they touch and are not translocated through plant's vascular system to other portions of the plant. In contrast, systemic herbicides are taken up by roots or foliage and translocated throughout the plant, so they can affect parts of the plant that were not directly exposed to the herbicide (Fig. 11.7). Contact herbicides kill plants quickly, often within a few hours of application, whereas systemic herbicides kill slowly and usually require several days or sometimes a few weeks to be fully translocated and effective. Contact herbicides require adequate spray volumes and complete coverage of the weed foliage for effective control, and repeat applications are usually needed to kill regrowth from underground parts of the plant. In contrast, systemic herbicides do not require complete coverage of weed foliage; if any portion of the weed receives the spray, herbicide molecules are translocated throughout the plant.

*Chemical control of broad-leaf weeds*

The selective post-emergence herbicides most commonly used to control broad-leaf weeds in the tropics include: 2,4D, 2,4DP (dichlorprop), MCPP (mecoprop), dicamba and triclopyr. Many different commercial formula-

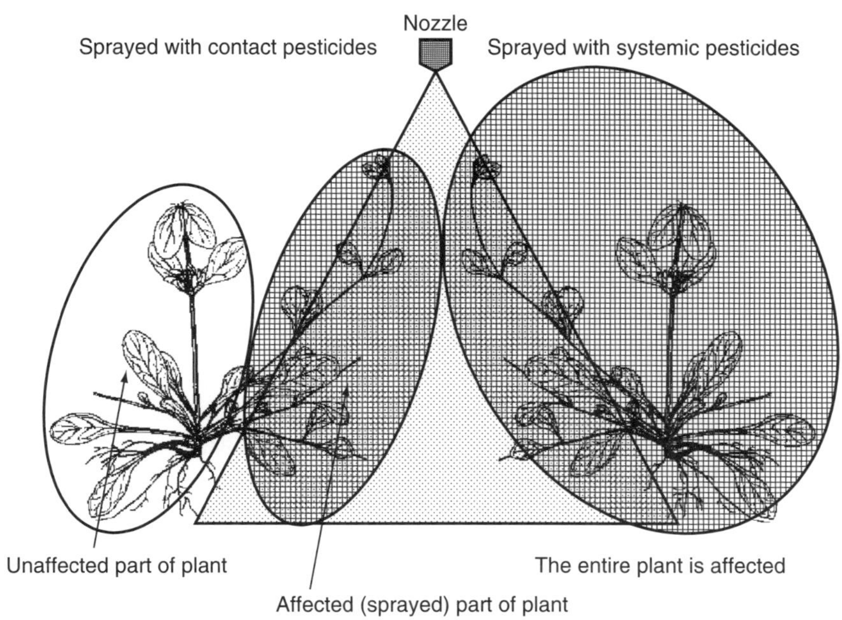


**Fig. 11.7.** Contact pesticides (herbicides, fungicides, insecticides) affect only those parts of the plant that they actually touch. Systemic pesticides can affect parts of the plant that were not directly exposed.

tions and mixtures of these compounds are available, but they are readily available in most countries. All of them are relatively harmless to turf-grasses when applied according to label recommendation, but all can damage trees, shrubs and flowers. Trees and shrubs are particularly sensitive to dicamba because this herbicide is mobile in the soil and can easily be taken up by roots.

The most effective control of broad-leaf weeds is obtained when post-emergence herbicides are applied as sprays to foliage. They must be applied when rain is not expected. Most current formulations require several hours for adequate foliar absorption, and if rain comes earlier, control may be poor. Once broad-leaf weeds are eradicated, properly conducted cultural practices, especially frequent mowing and adequate fertilization, can prevent their re-establishment for many years.

*Chemical control of grass and grasslike weeds*

The most troublesome grassy weeds in the tropics are crab-grass (*Digitaria* spp.), goose-grass (*Eleusine indica*) and nutsedge (*Cyperus* spp.). *Digitaria* is relatively easy to control in cynodon turf, especially when mowing height exceeds 10 mm. A chemical called methanearsonate (MSMA), a member of the herbicide family called organic arsenicals, is especially effective against *Digitaria* spp. Application of MSMA at the label-recommended rate usually produces excellent control and only minor, short-duration discoloration of cynodon turf. Goose-grass can be effectively controlled with a combination of more than one organic arsenical. Particularly effective is MSMA mixed with metribuzin, but the full label-recommended rate can be safely applied only on cynodon turf. Zoysia-grass turf is more sensitive to organic arsenicals, but when they are applied at reduced rates, control of *Digitaria* and *Eleusine* by MSMA can be moderately successful and zoysia injury only modest. Other species, such as paspalum, centipede grass, Bahia grass and St Augustine grass may be severely injured by organic arsenicals, so MSMA should be strictly reserved for cynodon and possibly zoysia grass, if modest discoloration of zoysia turf can be tolerated.

Fenoxaprop-ethyl, a member of the aryl-oxy-phenoxy herbicide family, produces good control of grassy weeds, especially *Digitaria* spp. in zoysia-grass turf. Cynodon and other turf-grasses do not tolerate this chemical and sustain severe injury, whereas zoysia grass shows only minor discoloration.

Sethoxydim is another aryl-oxy-phenoxy herbicide and the only one that provides excellent control of grassy weeds in centipede grass turf, but it cannot be used on any other turf-grass.

Paspalum, St Augustine grass and Bahia grass tolerate post-emergence herbicides poorly, and grassy weeds cannot be safely controlled in these species by chemical means.

Some weeds cannot be controlled by selective herbicides in particular types of turf. Spot treatment with a non-selective herbicide such as glyphosate is the most reliable means of removing these weeds from the turf. Foliage-applied glyphosate is translocated to roots, stolons and rhizomes and slowly kills the entire plant. Glyphosate should be sprayed on actively growing weeds, and wind drift toward non-target areas should be avoided.

Besides crab-grass and goose-grass, nutsedge (*Cyperus* spp.) seems to be the most troublesome weed in tropical turfs. It resembles grass, but like all sedges, it has a triangular stem with three-ranked leaves, not the round stem with two-ranked leaves typical of grasses. The *Cyperus* root system is fibrous and produces deep-rooted tubers or nutlets. Like all sedges, it typically thrives in soils that remain wet for extended periods of time. Historically, post-emergence chemical control of *Cyprus* was attempted with repeat applications of 2,4D, the organic arsenicals or a combination of the two, but extensive damage to most turf-grass species was a major concern. Substantial progress in *Cyperus* control was achieved in the mid-1990s when halosulfuron appeared on the market. Halosulfuron belongs to the sulfonyl-urea herbicide family and provides good post-emergence control of most sedges with only slight to moderate discoloration of major turf species.

Starting in the 1990s, substantial amounts of research have been conducted to develop genetically engineered turf-grasses resistant to non-selective herbicides. Just like genetically engineered agricultural crops, several species of turf-grasses have been modified to tolerate otherwise lethal doses of glyphosate. When such turf is sprayed with non-selective glyphosate, only genetically modified plants survive. This type of weed control allows eradication of all weeds as well as accidentally introduced turf-grass plants that may be of the same species but not genetically engineered. The resulting turf is absolutely weed free and extremely uniform. Among several genetically engineered species, St Augustine grass is presently being tested for introduction in the near future, and research on several other species is highly advanced.

Genetically modified turf-grasses will probably revolutionize turf-grass weed control, making it simple and safe. As in agronomic crops, weed control may depend upon only one chemical. Presently the biotechnological industries are concentrating their effort on glyphosate, one of the least environmentally harmful herbicides ever discovered.

## Diseases

Fungi, and sometimes nematodes, bacteria, viruses or adverse environmental conditions cause most turf-grass diseases. Literally hundreds of pathogens cause plant diseases, which vary widely according to geographical location and climatic conditions. Specific information about, and identification of, particular diseases and methods for their control should be sought from local plant-protection services. The discussion here will be limited to basic concepts of how a disease develops, management practices that can reduce disease occurrence, and the major methods used to eradicate diseases from infected turf.

Turf-grass diseases develop only when pathogen, susceptible host and suitable environmental conditions are present at the same time. Disease will not develop if only two of the three factors are present. For example if the pathogen and conditions favouring its development are present, but the turf-grass is tolerant to the pathogen, disease will not occur. Similarly, disease will not develop if the turf-grass is susceptible to a pathogen but the environmental conditions are not favourable (Fig. 11.8).

## Disease diagnosis

Before any control is attempted, the disease must be properly diagnosed. Although diseases are sometimes responsible for poor turf quality, they may not be the only cause; other factors, such as insects, drought, excessive water or fertilizer burn can cause symptoms that may be confused with signs of disease. Some turf-grass diseases are very typical and easy to identify, but others are not. Some disease-causing fungi produce abundant powdery spores or weblike mycelia, and sometimes obvious and

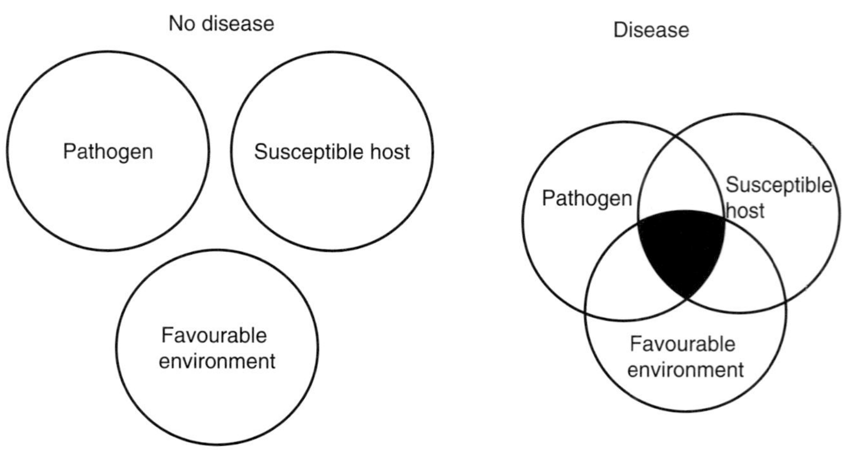


**Fig. 11.8.** Turf-grass diseases develop only when pathogen, susceptible host and favourable environmental conditions are all present at the same time.

typical spots or lesions appear on the individual blades, but in other cases, the entire leaf blades are affected. Sometimes symptoms are typical from the beginning, but sometimes they are overlooked because frequent mowing tends to destroy early evidence of the disease and only more severe, advanced symptoms are noticed.

If the turf-grass manager has difficulty with identification of a particular disease, the local plant-protection service or plant diagnostic clinic should be contacted. Local diagnosticians and plant pathologists are often most knowledgeable about the current disease problems in their own areas and can often help to diagnose diseases from descriptions of the symptoms. If not, or if the problem is severe or persists for a long time, turf samples should be delivered to the appropriate laboratory. Samples should include both healthy and unhealthy turf, should be taken at an early stage of disease development, should be about 10 cm in diameter, and should include at least 5 cm of roots below the thatch. The turf samples should be accompanied by information about recent weather conditions, cultural practices, fertilization and pesticide applications; photographs might also be provided.

If nematodes are the suspected cause of turf disease, their presence or absence must be confirmed before pesticides are used. Because nematicides are profoundly toxic, their application should be avoided unless it is essential, so nematode presence and count must be determined through accurate laboratory diagnosis.

In the tropics, disease management is generally less challenging than management of weeds or insects. Warm-season turf-grasses are much more disease resistant than cool-season species and, if they are infected, damage is usually less profound and relatively easier to control. In the arid and semi-arid tropics, fungal diseases are rarely a major problem. In the humid tropics, they are more common and usually develop in areas subject to additional stresses. Diseases caused by nematodes are relatively rare but much more common than in temperate climates. Nematodes are very plentiful in the soil. For the most part, they are harmless and feed mainly on fungi or bacteria, but some species can infect turf-grass roots, impair growth of the plants, and make them look chlorotic and wilted. Severity of disease usually depends upon the number of parasitic nematodes present in the soil.

### Methods of disease control

Like weeds, diseases in turf can be controlled by means of prevention, development of resistant cultivars, proper cultural practices and chemical interventions. The principal means of prevention is minimizing introduction of pathogens into the turf area. Sanitation of seeds, vegetative planting material and soil may reduce or eliminate the population of certain pathogens in the seedbed. Preventive measures are quite effective in reducing occurrence of diseases caused by nematodes, but elimination of numerous fungal pathogens by prevention alone is very difficult. Many fungi are easily spread by wind, equipment or animals, and turf usually harbours them permanently. An important aspect of disease prevention is development of turf-grasses less susceptible to pathogens. Numerous disease-resistant cultivars have been developed for temperate climates, where diseases are widespread and often devastating. Because warm-season species are naturally much more resistant, considerably less effort has been put into development of disease-resistant varieties. Among warm-season species, several cultivars resistant to the virus causing St Augustine grass decline (SAD) have been developed.

The most effective preventive measure is to provide unfavourable environmental conditions for the growth of pathogens. Some factors, like temperature, are quite difficult to modify in the tropics, but others, such as plant nutrition, soil moisture and duration of leaf wetness, can be altered.

Temperature has a direct effect on the growth rates of fungi, but unlike the situation in temperate climates, where temperature is very important, high temperature is not a primary factor promoting development of disease. In the tropics, humidity within the turf-grass canopy seems to be more important. Moisture-saturated air allows germination of most fungal spores and the growth of fungal mycelia. After sunset, humidity inside the turf canopy usually increases and reaches the dew point. The moisture then condenses in the form of little droplets, which remain on the turf until sunrise. When the humidity is high and evaporation of water from the leaf surfaces in the morning is slow, the favourable period for growth of fungal pathogen growth becomes longer. More than 12 h of leaf wetness are needed for most pathogens to grow and penetrate plant tissue. If actively growing fungal structures are dried rapidly before the fungus has penetrated the plant surface, the pathogen's life cycle is broken, and disease does not develop. Therefore, any measure that reduces the duration of leaf wetness decreases development of disease, and any measure increasing leaf wetness promotes it. In the morning, the period of leaf wetness can be reduced mechanically by dragging of ropes or water hoses across the grass to knock the dew off the blades, as well as by watering around the time of sunrise. Humidity in the canopy can also be reduced by removal of obstructions that reduce the movement of air. Spacing or pruning shrubs and trees in the landscape greatly improves air movement and speeds the evaporation of dew. On the other hand, late-afternoon watering extends the period of leaf wetness. Because the turf may not dry before evening dew is formed, fungus can start its growth cycle several hours earlier, and mycelium may penetrate the plants before they dry off in the morning. Afternoon or early-evening irrigation should be avoided whenever possible; turf-grasses should be watered at times that assure that the leaf surfaces will dry before nightfall.

The amount and timing of fertilization can greatly influence occurrence of turf-grass diseases. Typically, high levels of nitrogen increase the risk of disease, but low nitrogen levels promote some of them, such as the fungi *Sclerotonia homoeocarpa* on cynodon, zoysia grass and centipede grass and *Puccinia* spp. on zoysia grass. Numerous diseases may also become more severe when the plant suffers from deficiencies of nutrients, especially potassium, calcium, phosphorus and magnesium.

Mowing practices can also affect diseases. Mowing creates open cuts in the leaf cuticle, allowing pathogens an easy entrance into the mesophyll and allowing expansion within the plant. The risk of infection can be increased even more when leaf tips are severed by dull mower blades or when mowed turf-grass is wet and spores can readily move from infected to healthy plants. Excessively low mowing height can also increase disease occurrence by increasing stress on the turf-grass plants.

Usage of pesticides to control insects in turf-grasses may indirectly influence activity of pathogenic fungi. Insecticides are often toxic to saprophytic insects or other arthropods, and the disruption they cause to the ecological balance may weaken turf-grass plants. Herbicides may cause direct injury and therefore reduce plants' vigour and increase their susceptibility to pathogens.

### *Chemical disease control*

In tropical climates, sound management practices can prevent many diseases from occurring on the larger scale, but sometimes chemical intervention is needed to prevent excessive damage to high-value turf areas such as putting-greens, tees and some sport fields. Turf diseases can be controlled with fungicides, nematicides and fumigants. Fumigants are used very infrequently and only on small areas such as golf greens or tees, where diseases or nematodes have been present in the past and may cause severe and long-lasting problems in the future. Other turfs in tropical locations seldom undergo fumigation.

Nematicides are used more frequently but overall not often. They are drenched into the thatch and soil and kill nematodes that come in contact with the chemical. Besides nematodes they also kill numerous insects and other soil fauna such as earthworms.

Fungicides, which kill or inhibit growth of fungi, are used quite commonly. They should, if possible, be applied at the earliest stage of the disease. This type of control, called *curative*, restrains the pathogen after a mild infection has already become established, but turf areas that have a history of developing a particular disease under certain weather conditions are often sprayed before disease development. This type of control is called *preventive*; it restrains the pathogen before it enters the plant. Because of environmental concerns, continuous preventive disease control is highly discouraged, but it is nevertheless still practised on some golf courses, where broad-spectrum fungicides are applied throughout the year.

Fungicides that protect turf-grass against infection can be grouped into two types: *contact* and *systemic*. Contact (non-systemic) fungicides kill spores and mycelium on the surface of the turf-grass plant. Because turf is subject to mowing as well as wash-off by irrigation and rainfall water, contact fungicides are usually effective for no more than several days. In contrast, systemic fungicides are applied mostly to the soil and are then absorbed and translocated throughout the plant vascular system, where they kill growing fungus as it penetrates the plant. Because of their residual activity and internal action, systemic fungicides are effective for much longer than contact fungicides, often for several weeks. Systemic fungicides are usually formulated for application as a spray or granules. Because many of them are absorbed primarily by the root system, liquid forms should be applied in fairly large volumes of water or watered in immediately after spray application. Granular forms should be applied prior to irrigation or rainfall. On the other hand, all contact fungicides are formulated to be sprayed. The volume of water is typically higher than for herbicides and ranges from 10 to 15 l of water per 100 $m^2$ (1000–1500 l/ha).

Disease-control programmes should never depend upon fungicides alone. Disease-management procedures are meant to complement rather than displace one another. Selection of proper cultivars, best management practices and occasional use of chemicals guarantee healthy and vigorous turf.

## Insects

Insects are the most abundant animals living on earth. Most of them decompose organic matter, and the large majority are entirely harmless. Some, however, are significant pests of plants and animals.

The usual description of insects as six-legged, segmented creatures with their skeletons on the outsides rather than the insides of their bodies actually applies most often to adult insects, which in many cases live only to reproduce, do not feed and cause no damage. During their life cycle, insects undergo a series of changes called *metamorphosis*. For some insects, metamorphosis is complex, and for others it is simple (Fig. 11.9).

The former begin life as eggs, which hatch into larvae, which in turn become pupae, and finally reach the adult form. In insects with simple metamorphosis, the eggs hatch into nymphs that resemble adults but are

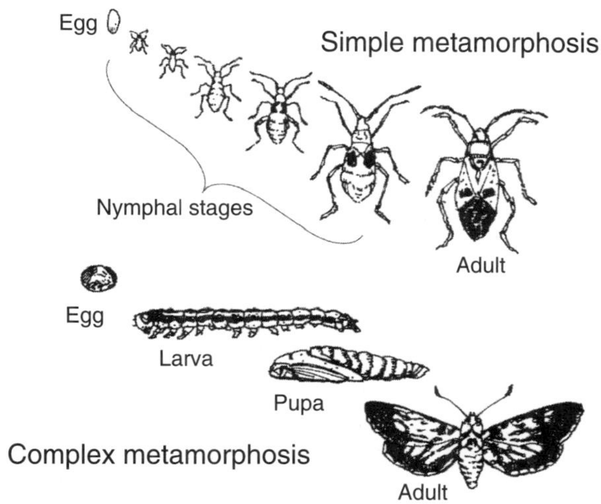


**Fig. 11.9.** In simple metamorphosis, the eggs hatch into nymphs that resemble adults. In complex metamorphosis, eggs hatch into larvae, which then become pupae, which in turn metamorphose into the adult form.

smaller and not fully developed. Most insect pests of turf undergo full metamorphosis, and the larvae rather than the adults cause the damage. In those that undergo simple metamorphosis, the nymphs usually feed as they grow.

Some insects feed by chewing and others by sucking. Chewing insects ingest plant tissues or organic residues directly. Sucking insects use piercing mouthparts to poke through the surfaces of leaves, stems or roots and suck the plant fluids.

Insects presenting a direct threat to the turf-grass community are root-feeding insects such as grubs, billbugs and mole-crickets as well as shoot-feeding insects like web-worms, army-worms and chinch-bugs (Fig. 11.10). Burrowing insects such as ants and cicadas are considered a nuisance but do not cause direct damage.

Unlike weeds or diseases, insects are not very effectively controlled by preventive measures or cultural practices. In fact many insects are more likely to attack well-maintained turf because healthy plants with lots of fresh growth provide a plentiful source of food. To manage insect pests effectively, turf managers must have a working knowledge of their life cycles and habits in their feeding areas.

## Insect detection

Before any control is attempted, the presence of insects must be detected. Accurate diagnosis of insect infestation requires periodic visual observa-

**Fig. 11.10.** Leaf- and shoot-feeding larva of sod web-worm.

tion of the turf and other parts of the landscape. For example, the population of grubs can be estimated from the number of adults feeding on host plants or from the presence of beak holes in the sod produced by birds searching for larvae. Another example is presence of moths flying in front of shrubs shortly after sunset, a good indication of approaching infestation with sod web-worms.

Pests are never evenly distributed throughout the turf; rather, they occur in isolated spots or pockets. Frequently, certain areas, such as moister low spots, elevated dry spots or windbreak areas are more vulnerable to certain insects than others. As a result, these areas should be inspected first and more thoroughly. Turf thinning, dead or dying patches, chewed or frayed grass blades, and distinctive webbing or faecal pellets are potential evidence of insect pests. Still, the best indication of an insect infestation is physical presence of the insect.

Numerous sampling techniques are utilized to look for insects infesting turf. For example, a floating technique is used to sample arthropods, especially chinch-bugs, *Blissus* spp. One end of a plastic cylinder or large can (15–20 cm in diameter), open at both ends, is forced into the turf surface and filled with water. Because water continuously escapes, water must be added for 5–10 min, which is usually sufficient for chinch-bugs to float to the surface, be recognized and be counted (Fig. 11.11). Another sampling technique involves use of 20–40 ml of laundry detergent mixed with 5–10 l of water. Minutes after application of this solution to turf, caterpillar-type larvae emerge to the surface, where they can be identified and counted. The most destructive method, soil sampling, is sometimes necessary for confirming the presence of soil-inhabiting insects such as

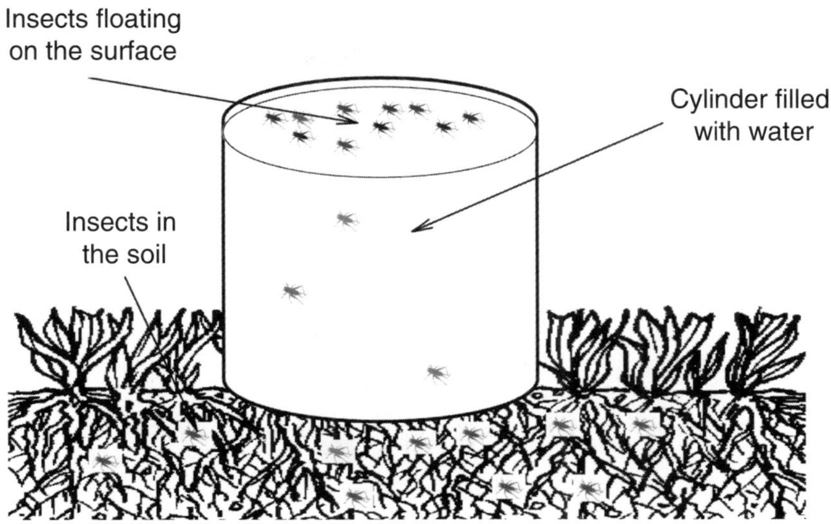


**Fig. 11.11.** A floatation technique is used to sample arthropods, especially chinch-bugs.

white grubs. A square area of sod (25 × 25 cm) should be cut to a depth 1–2 cm below the root zone and lifted up. The roots and soil beneath should be inspected for presence of grubs before being replaced.

## Methods of insect control

Cultural practices such as proper mowing height, fertilization, soil aeration and watering do not prevent infestation but can help the turf to resist higher pest populations without serious damage. Some insect-resistant varieties of turf-grasses have been released and are available on the market. Among warm-season species, for example, development of a chinch-bug-resistant variety of St Augustine grass was a significant breakthrough in management of this particular insect.

Chemical controls, however, are usually the most reliable means of controlling properly identified insects and, even if they cannot eliminate all of the insects, may reduce their numbers to a manageable level. The timing of insecticide application is one of the most critical aspects of pest control. Insects are more susceptible to chemicals at some stages of life than at others. Ordinarily, the resting stages, such as eggs and pupae, are unaffected, whereas active stages such as larvae, nymphs and sometimes adults are susceptible to chemicals. The manager's goal is to apply chemicals at the time of peak susceptibility. Placement of the insecticide is also an important component. Most insecticides kill only by direct contact with the insect, so bringing insects closer to the surface may be important. Extensive irrigation usually makes insects crawl toward the soil surface. Root-feeding insects must be treated with insecticides that move downward through the soil to the place of the greatest insect activity. The best results are obtained when the two movements are combined: when insecticides penetrate the soil and at the same time insects move upward toward its greatest concentration. This combination would normally require at least 10–15 mm of irrigation water after application of the insecticide. Watering should follow insecticide application quickly, because in addition to encouraging increased contact with insects, immediate drenching reduces insecticide losses associated with foliar adherence, volatilization and photodecomposition.

On the other hand, shoot-feeding insects ingest mainly insecticide that adheres to the foliage. If these insects are the main targets, irrigation should be delayed as long as possible after application of the insecticide. Because many shoot-feeding insects feed at night, the insecticide should be applied late in the day.

Insect control is a complex issue requiring extensive knowledge of both biology of insects and chemistry of insecticides. Turf-grass managers in a particular tropical location are usually confronted with only a few species of insects that seriously damage turf. A discussion of all harmful tropical insects and methods for their control is beyond the scope of this book. Up-to-date information on particular pests in specific locations is frequently released by research institutions and chemical companies. It is typically published in professional magazines as well as on the Internet, two excellent sources of information that should be used to the fullest advantage.

# Appendix
# Application of Fertilizers and Other Dry Materials with a Spreader

Greg Wiecko, UOG Station/CNAS, Mangilao, Guam 96923, USA
telephone (671) 735–2132, fax (671) [734–7657], e-mail gwiecko@uog9.uog.edu

Dry fertilizers can be applied with either a drop (gravity) spreader or a rotary (centrifugal) spreader. A drop-type spreader has the advantage of applying fertilizer in a quite accurate pattern, but it requires that each edge of successive 'passes' (swaths of area covered by the spreader) meet exactly; if they do not, 'skips' (unfertilized areas between passes) and overlaps will be noticeable. The rotary spreader generally has a wider pattern of distribution than a drop spreader and can therefore cover a larger area in the same amount of time. Its application pattern also gradually diminishes with distance from the machine, reducing the probability of skips. This spreader is harder to calibrate and application is less precise, but it is sufficiently accurate for most home or lower-maintenance areas.

One method that increases uniformity of application is to divide the material to be applied (see below for calculation of the proper amount) into two equal portions applied separately. Use a spreader calibration that will deliver one-half of the desired amount of material, and apply one-half of the material over the entire area. Then turn the spreader direction 90° from that of the initial application and apply the remainder

**Fig. A.1.** Drop-type spreader.

over the entire area. This process effectively eliminates skips in the coverage.

Spreader calibration is the measurement of the output as the spreader is operated over a known area, so as to determine what settings will ensure the desired rate of application.

## Calibrating a Drop Spreader

Follow these steps in order to calibrate a drop spreader (Fig. A.1):

1. Check the spreader to be certain that it functions properly.
2. Measure the width of the spreader opening and the circumference of one of its wheels. For example, the spreader in Fig. A.1 is 60 cm wide, and the wheels have a circumference of 1 m. Mark a wheel with a dot to provide a starting point. When the wheel has been turned 10 times (rolled 10 m), you know that the 60-cm-wide spreader has covered an area of 6 $m^2$ (0.6 m $\times$ 10 m).

3. Fill the spreader with the material you wish to apply (fertilizer, seed, etc.). Fill the hopper only to the level you will use when the material is actually applied.
4. Make several short trial runs over the area and practise opening and closing the spreader. Do so gradually, not in a fast, jerky way. Make sure the material drops from the full length of the opening.
5. Determine the weight of the fertilizer applied by the spreader over a known area: some, but not all, commercial spreaders come equipped with an attachable catch pan that fits under the spreader opening. When no catch pan is available, this calibration requires two people: one to hold the spreader and squeeze the opening lever and the other to lift one wheel and turn it manually. While the first person holds the lever in the open position, the other should turn the wheel ten times. Collect the fertilizer that drops from the spreader opening on a large piece of paper and weigh it.
6. If the amount you collect in the catch pan is less than that intended, increase the gate opening. If it is more, reduce the opening. Repeat this procedure until the amount dropped is within 10% of the desired amount (110–130 g, if the intended amount is 120 g). After reaching this final setting, repeat the procedure at least one more time to be sure that the calibration is correct.

The same calibration procedure is used for fertilizer, seeds, granular pesticides or any material to be applied, but remember that the rate of application depends on the physical properties of the material, so the same settings cannot be used for different materials.

## Calibrating a Rotary Spreader

Follow these steps in order to calibrate a rotary spreader (Fig. A.2):

1. Check the spreader to make certain all parts are operating properly. Then follow steps 2 and 3 to determine the 'effective' width of application.
2. Fill the hopper about half full with the material you plan to apply and set the spreader on 'medium', i.e. about half open. Run the spreader over bare ground or a hard surface, where the width of the surface covered by the spread material can be measured.
3. The amount of material applied by a rotary spreader (Fig. A.2) is not constant across the entire width of application. More material is applied towards the centre and less at the edges. If the full application width is 5 m, only about the central 3 m receives an approximately even application. The outer 1 m on each edge receives much less. The central area is called the effective width of application, and it (not the full width) is used in all calculations.

**Fig. A.2.** Rotary-type spreader.

4. Mark off a test distance that, when multiplied by the effective width, equals 100 $m^2$. For example, an effective width of 3 m multiplied by a distance of 33.3 m yields an area of about 100 $m^2$.
5. Fill the hopper with a known weight of fertilizer and adjust the spreader to the lowest setting that will allow the material to flow. Push the spreader along the calculated test distance, opening the hopper at the starting line and closing it at the finish. Weigh the material left in the spreader and subtract the result from the starting weight. The difference is the amount applied per 100 $m^2$ at that spreader setting.
6. Repeat step 5 at successively higher settings (greater openings) and record the amount of material applied at each setting.
7. Select the spreader setting that most closely approximates the desired rate of application, and set the spreader accordingly. To obtain uniform spread of material, remember to set the spreader at half the

desired rate of application and make two passes at 90° to each other. To achieve the proper application overlap, use the effective width as the width of each pass. For example, if the effective width is 3 m, then after each pass, move the spreader over 3 m from the centre of the tyre tracks of the previous pass, to ensure a fairly constant rate of application over the entire area.

Again, remember that the spreader must be calibrated for each type of material to be applied. The same settings will not be correct for all materials.

## Determining the Amount of Material to be Applied

Fertilizer, seed and pesticide labels usually list the recommended application rate in kilograms per hectare (g/100 $m^2$). If a fertilizer label recommends 2 kg/100 $m^2$, for example, and the area to be fertilized is 6 $m^2$, the amount of fertilizer needed is 120 g. The calculation is demonstrated below. $X$ represents the quantity to be applied per 6 $m^2$.

### Calculating from a Proportion

$$\frac{X\mathrm{g}}{6\,\mathrm{m}^2} = \frac{2000\,\mathrm{g}}{100\,\mathrm{m}^2}$$

Cross-multiply (i.e. multiply both sides of the equation first by the denominator on the left-hand side, then by the denominator on the right-hand side) to obtain

$$X\,\mathrm{g} \times 100\,\mathrm{m}^2 = 2000\,\mathrm{g} \times 6\,\mathrm{m}^2$$

Then divide both sides by 100 $m^2$ to obtain

$$X = 120\,\mathrm{g}$$

## Calculations of Turf Areas

Some turfs have shapes of distinct geometric figures and can be measured with high accuracy. Some others may be irregular and measurements can be only approximated (Fig. A.3). Superior accuracy is

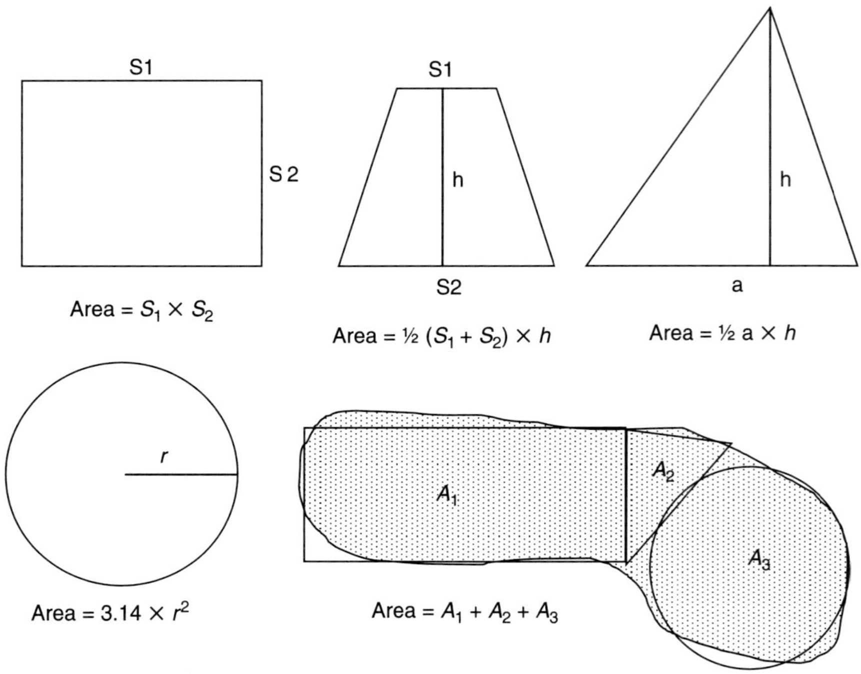


**Fig. A.3.** Formulae for calculating areas of common geometric configurations.

usually not required for calculating fertilizer seed or pesticide applications. Irregular shapes can be simplified, individual areas calculated and totalled.

# Glossary

**Acidic:** Having a pH of less than 7.0.

**Active ingredient:** In a pesticide formulation, the material that actually destroys the target pest or performs the desired function.

**Aeration:** The exchange of gases between the soil and the atmosphere.

**Aerification:** The process of disturbing the soil by mechanical means to relieve compaction and maximize air, water and nutrient availability to turf-grass.

**Aerification, hollow-tine:** The creation of vertical channels in the soil by means of hollow cylinders that physically remove cores of turf and soil to a specified depth; also referred to as 'core aeration'.

**Aerification, solid-tine:** The creation of vertical channels in the soil by displacement, without physical removal of cores.

**Aggregate:** A soil structure made up of many soil particles held together in a single mass or mineral material of uniform fine size used in construction projects.

**Alkaline:** Having a pH higher than 7.0.

**Annual:** A plant that completes its life cycle within one growing season.

**Application, post-emergent:** Application of herbicides, to plants that have already germinated and emerged from the soil.

**Application, pre-emergent:** Application of herbicides, designed to prevent the germination of plant seeds.

**Bacteria:** Microscopic, single-cell organisms, some beneficial and some pathogenic, that live in a variety of forms throughout the biosphere.

**Burning, of leaves by fertilizers:** Desiccation of leaves due to application of fertilizers or other materials in excessive concentrations.

**Calcium carbonate:** A compound used to raise soil pH, but rarely on established turf, because of its burn potential.

**Calibration:** The process of adjusting equipment to ensure the desired rate of application of chemicals.

**Canopy:** The continuous layer formed by the leaves of a stand of turf-grass.

**Cation:** A positively charged ion.

**Cation exchange capacity (CEC):** The total quantity of exchangeable cations that a soil can adsorb.

**Chelate:** A chemical formulation in which a metal atom (very often a micronutrient such as iron) is bound with an organic component to improve the overall uptake of the micronutrient.

**Clay:** A soil material with particles of 0.002 mm diameter or less, noted for high moisture- and nutrient-holding capacity; also used to refer to a soil mixture containing more than 40% clay.

**Coated fertilizer:** Granular fertilizer that has been coated with a material of known permeability to ensure the controlled release of nutrients into the soil.

**Compaction:** Compression of the topsoil, primarily due to foot or vehicular traffic and tending to prevent the passage of air, water and nutrients into the soil.

**Contact herbicide:** A pesticide that acts only on the portion of the plant it touches.

**Controlled-release fertilizer:** A product engineered to release nutrients into the soil over a designated period of time.

**Crown:** A group of nodes separated by very short internodes in the central core of a turf-grass plant or the elevated centre portion of a sports field, raised to promote the runoff of surface water.

**Cultivar:** A variety or subdivision of a plant species that, on the basis of morphology or performance characteristics, can be distinguished from other plants within that species.

**Curative control:** Application of a pesticide after the outbreak of disease or infestation, as opposed to preventive application.

**Desiccation:** The withering of plant tissues due to acute lack of moisture.

**Divot:** A piece of turf torn up by a golf club as it strikes the ball.

**Divot opening:** The scar on the turf surface that results from formation of a divot.

**Drain, French:** Strictly speaking, drainage trench filled with stone or gravel, containing no pipe.

**Emulsifiable concentrate:** A pesticide formulation that consists of one liquid suspended in another, usually an oil-based pesticide carrier suspended in water.

**Evapotranspiration:** The loss of water from the soil through a combination of evaporation from the surface and transpiration through plants.

**Fertilizer analysis:** Percentages by weight of nitrogen, phosphate and potash in a fertilizer product.

**Fertilizer, natural organic:** A nutrient source of plant or animal origin, typically with low nutrient analysis and requiring soil microbial activity to convert nutrients to plant-available forms.

**Fertilizer, synthetic organic:** A chemically engineered carbon-based nutrient source.

**Field capacity:** The upper limit of storable water in a field layer after the water has drained through as a result of gravity.

**Foliar absorption:** Uptake of nutrients through the stomata and cracks in the leaf cuticle.

**Foliar feeding:** Application of nutrients to the leaves, for foliar absorption.

**Fumigant:** A gaseous pesticide.

**Fungicide:** A pesticide used to control fungi.

**Growing points:** The portions of a plant where new tissues form; also known as meristematic zones.

**Herbicide:** A pesticide used to control unwanted vegetation.

**Herbicide, non-selective:** A herbicide that kills all plant tissues that it touches.

**Herbicide, post-emergent:** A herbicide that is applied to control weeds after their germination and emergence from the soil.

**Herbicide, pre-emergent:** A herbicide that is applied to turf to control germinating weed seeds.

**Herbicide, selective:** A herbicide that affects only a specific weed or group of weeds and has minimal effects on desirable plants.

**Hybrid:** The offspring of parents from different species or varieties.

**Hydrophobic:** Tending to repel water.

**Infiltration:** The downward entry of water into the soil.

**Insecticide:** A pesticide used to control insects.

**Irrigation system:** A system for delivery of supplementary water to plants, often a network of underground pipes and pop-up sprinklers controlled by manual or automatic valves.

**$LD_{50}$:** The dosage of a toxic substance that kills half of the test organisms in an acute study.

**Loam:** A type of soil composed of moderate amounts of sand, silt and clay.

**Macronutrients:** The nine nutrients required by plants in large quantities: nitrogen, phosphorus, potassium, calcium, magnesium, sulphur, carbon, hydrogen and oxygen.

**Macropores:** Large pores in soils that drain freely by gravity.

**Micronutrients:** The seven nutrients required by plants in small quantities: manganese, iron, boron, zinc, copper, molybdenum and chlorine.

**Micropores:** Small soil pores that retain water by capillary action and do not drain freely by gravity.

**Monostand:** An area of turf-grass made up of one species of plant.

**Mower, flail:** A mower with pivoting blades or flails that spin horizontally at high speed around a vertical axis.

**Mower, reel:** A mower that shears the grass between a blade on a spinning cylinder and a bed knife at the base of the cutting unit.

**Mower, rotary:** A mower that cuts by the impact of a horizontal spinning blade.

**Mower, vertical (verti-cutter):** A device with blades that cut vertically into the turf canopy to sever lateral stems, thin grass and dethatch.

**Nematicide:** A pesticide used to control nematodes.

**Nematodes:** A group of soil-borne, microscopic, worms, some of which attack the root systems of plants.

**One-third rule:** The mowing rule that specifies that no more than one-third of the total leaf area be removed at any one cutting.

**Overseeding:** The application of additional turf-grass seed to existing turf.

**Peat moss:** A commonly-used organic amendment consisting largely of undecomposed (or only slightly decomposed) organic matter which accumulates under conditions of excessive moisture.

**Perched water table:** Layering of soil designed to retain moisture in a surface layer above a coarser and better-drained lower layer.

**Percolation:** The downward movement of water through the soil.

**Perennial:** A plant that can reproduce several times over several growing seasons.

**Pest:** In turf-grass management, any living organism that competes with turf-grass plants for nutrients, light, water, air or space, or that otherwise adversely affects turf-grass performance, for example by causing disease.

**Pesticide:** A chemical agent used to kill pests.

**Plant growth regulator (PGR):** A chemical agent applied to plants to suppress their growth.

**Plugging:** The process of installing turf-grass by planting rooted pieces or 'plugs' of live turf, which will be encouraged to spread over a desired area.

**Polystand:** An area of turf-grass made up of more than one species of plant.

**Porosity, soil:** The amount of air space between the particles of soil, expressed as a percentage of the volume of soil not occupied by solid particles.

**Pregermination:** A method used to reduce seed germination time in which seed is soaked or placed in a moist environment to encourage partial germination, then planted as soon as possible to prevent desiccation.

**Preventive control:** Application of pesticides before the outbreak of disease or an infestation, usually on turf that has a history of such outbreaks or infestation.

**Rhizome:** An underground, horizontally growing plant stem that produces both upwardly growing shoots and new roots from nodes; distinguished from a root in that it has buds and nodes.

**Root zone:** The layer of soil in which the roots of turf-grass plants are found and from which they must draw air, water and nutrients.

**Silt:** A soil material made up primarily of microscopic soil particles ranging in size from 0.002 to 0.05 mm.

**Slicing, soil:** The process of cutting vertically into the soil to sever horizontal stems and promote aeration.

**Sodding:** The process of installing mature turf in large or small sheets, as opposed to spreading seed or sprigs over a prepared area of bare soil.

**Soil structure:** The combination or arrangement of soil particles into aggregates.

**Soil textural triangle:** A triangle diagram illustrating the range of particle sizes for the 12 textural classes of soil.

**Soil texture:** The relative coarseness or fineness of a soil, determined by the relative proportions of sand, silt and clay particles.

**Spiking, soil:** The process of using equipment with blade-like protrusions to puncture vertically into the soil, severing horizontal stems and promoting aeration; differs from slicing in that spiking punctures are not continuous openings in the soil like the slices created by slicing equipment.

**Spreader, centrifugal:** A rotary device used to apply seed, fertilizer and other granular materials; also called a rotary spreader.

**Spreader, drop:** An agricultural implement used to spread granular material over an area by allowing it to fall through openings of a specific size.

**Sprigging:** The process of vegetatively establishing turf-grass by spreading rhizomes or stolons over a prepared seed bed and pressing them into the soil.

**Stolons:** Creeping, above-ground stems that take root at the nodes to form new plants.

**Stresses, turf-grass:** Force that tends to strain or damage turf-grass, like drought, extreme temperatures, pests and the mechanics of player traffic.

**Surfactant:** A chemical additive that improves spreading, dispersing and/or wetting properties.

**Syringing:** The process of lightly watering turf, mainly to cool the turf canopy.

**Thatch:** An intermingled layer of living and dead grass stems, roots, and other organic matter found between the soil surface and the grass blades.

**Tissue test:** The chemical analysis of plant leaves to determine the level of nutrients present in the plants.

**Top-dressing:** The addition of sand or soil to the surface of the turf to level the turf surface and promote thatch decomposition.

**Vegetative planting:** Installation of turf-grass by distribution of live plant material, such as sprigging plugging or sodding.

**Viruses:** Disease-causing sub-microscopic organisms.

**Volatilization:** Conversion from a liquid to a gas.

**Warm-season grasses:** Grass species adapted to a climatic area where winter is relatively mild and daytime temperatures during the growing season are usually above 25–28°C.

# Index